The Politics of Climate Change Metaphors in the U.S. Discourse

Othman Khalid Al-Shboul

The Politics of Climate Change Metaphors in the U.S. Discourse

Conceptual Metaphor Theory and Analysis from an Ecolinguistics and Critical Discourse Analysis Perspective

palgrave
macmillan

Othman Khalid Al-Shboul
English Language and Translation
Jadara University
Irbid, Jordan

ISBN 978-3-031-19015-5 ISBN 978-3-031-19016-2 (eBook)
https://doi.org/10.1007/978-3-031-19016-2

Cover illustration: David Robertson / Alamy Stock Photo

This Palgrave Macmillan imprint is published by the registered company Springer Nature Switzerland AG.
The registered company address is: Gewerbestrasse 11, 6330 Cham, Switzerland

To the soul of my father Khalid Al-Shboul who put me on the road to success in my life, I dedicate this work to him.

Preface

I think that metaphor does have powerful impacts not only on our opinion (how we think of the social world) but also on how we react and act accordingly, in a specific manner. My motivation behind writing this book stems from my belief that metaphor can even change more than we think. I remember when I first studied metaphor at school. It was all about how we add beauty to what we speak and write. But at the graduate level, I looked at a great deal of research that examines metaphor as a cognitive tool.

This book develops theories and methodologies from critical discourse analysis (CDA) and ecolinguistics to explore the role of metaphor in the context of political climate change discourse in the U.S. This book is thus a bridge between critical discourse analysis (CDA) and ecolinguistics through metaphor. The book aims to make a contribution to metaphor (theory and analysis) and climate change studies, guided by the principles of CDA and ecolinguistics. It is an attempt to further explore how our thought can be shaped about environmental issues when metaphor is employed in political contexts. This book helps students and researchers in cognitive linguistics and related fields develop sufficient knowledge of metaphor theory and analysis from a cognitive perspective. It enriches CDA and ecolinguistics with cognitive aspects of analysis.

The book contributes to knowledge in three main areas: contribution to metaphor theory and analysis, contribution to studying climate change debate in the U.S. and analysis, and contribution to ecolinguistics research.

This book presents many examples of metaphors that help students connect theoretical concepts introduced in the book with how metaphor is processed. This book thus presents information at theory and analysis levels and thus the reader can be equipped with the necessary tools to understand the different aspects of metaphor.

Irbid, Jordan

Acknowledgment

I would like to express my deepest appreciation to the executive and production editors (Cathy Scott, Kishor Ramesh, Petra Treiber and Ulrike Stricker-Komba) from Palgrave Macmillan and Springer Nature, who worked on the book production. I want also to thank the greatest cover designer of this book. Thanks to everyone on my publishing team in Springer Nature for their editorial support and guidance.

I highly appreciate the anonymous reviewers from Palgrave Macmillan and Springer Nature for their careful reading of my manuscript and insightful comments and thoughtful suggestions during the various stages of the book publication. Their invaluable guidance and feedback helped me in completing this project successfully.

I would like to express my special thanks of gratitude to Dr. Sage Graham at the University of Memphis, USA, for helping me find a well-esteemed publisher for my book.

I am extremely grateful to all my family for their support. Without that support, this project could not have been completed. I am so grateful to my mother Badriah for her constant support and encouragement. She has never stopped to pray for my success.

I am also so grateful to my wife Sarah for her constant support. She has kept my motivation high during this journey. My gratitude is to my sons (Mohammad and Sanad O. Al-Shboul) who have been my motivation to move forward in my practical and academic life.

There are many people who helped me during the journey, bringing the book to fruition. I am so grateful to them all.

Last but not least, I would like to thank everyone who helped and motivated me to work successfully in my academic life.

Contents

List of Figures

List of Tables

1
Introduction

The first chapter introduces the topic of this book including how and why climate change is debated in the United States. It discusses how climate change is framed and moved from science into politics. As was the case with many issues such as gun control and abortion, this problem (climate change) transformed to and resituated in political contexts by politicians in the U.S., especially Republicans and Democrats. In addition, this chapter presents metaphor as an important feature that politicians draw on to shape realities about such environmental issues as climate change. Therefore, this explains, also discusses, why many researchers in linguistics and cognitive disciplines have studied this environmental issue. Additionally, this chapter introduces key terms, such as CDA, ecolinguistics, framing etc. which are used throughout this book.

Overview

This book outlines an extension to traditional conceptual metaphor analysis with a particular focus on the issue of climate change. This book aims to shed light on: 1. What are climate realities that politicians create

O. K. Al-Shboul, *The Politics of Climate Change Metaphors in the U.S. Discourse*,
https://doi.org/10.1007/978-3-031-19016-2_1

through metaphor? How can these realities be perceived by the public in the U.S? How can the addressee be directed according to these realities? How can these realities help frame climate change? How do politicians stress urgency? Which metaphors stress stronger urgency than others? This book is guided by the principles of CDA theory (how language is used by politicians?) and Ecolinguistics, based on assumptions drawn from these two approaches; that is, language is powerful, politicians have access to power resources that the public do not, and the issue (climate change) studied is related to the environment which is of top concern to many people in all over the world. Furthermore, this book investigates how climate change is framed in the U.S. and World media and, in a separate chapter, presents the authors' ecosophy, developing a model for ecosophy in climate change studies.

Below are significant and original contributions from the book:

First, I offered in-depth explanations of how the speaker makes realities, how the addressee perceives realities and how he/she can comply with the speaker's desires from a cognitive perspective. I, therefore, established metaphor opposition theory (positive and negative sides of a reality formed by a metaphor) and integrated this theory with the already established theory 'mapping theory' (Lakoff & Johnson, 1980) to build up a theoretical framework that helps researchers analyze metaphor from a cognitive perspective more deeply and comprehensively. In more detail, I combined the two theories (mapping and metaphor opposition) through 'feeding'. The term 'feeding' is used in phonology to mean "to open the door for the other rule (sound change) to apply". Accordingly, in my work, I argue that mapping feeds metaphor opposition (that is, it opens the door for the metaphor opposition to be activated). This book thus extends critical metaphor analysis (Johnathan Charteris-Black, 2018) and critical discourse analysis (Norman Fairclough, 2014) with this cognitive analysis for the purpose of offering deeper explanations of not only how the hearer perceives realities but also how he/she can comply with the speaker's desires (based on warnings and motivations he/she perceives through metaphor). That is, how the addressee may respond to the speaker's requests (e.g., to fight climate change due to its disastrous consequences) since the speaker implicates a warning of consequences and an encouragement (perceived) through metaphor. Thus, mapping and opposition schemas are provided to explain this from a cognitive perspective.

Second, this book explains how metaphor frames this environmental issue (climate change as a problem and solutions) and how it affects the addressee's perceptions of this issue and how he/she can respond. I found a formula (if, then) for the framing of climate change and how the addressee may perceive this framing in terms of a problem and procedures. The analysis also in this book shows the centrality of the journey metaphor in mixed metaphors. I found that almost all the mixed metaphors used by politicians arguing against the exit from the Paris agreement, in this book, consist of Journey and War, Journey and Cleanliness, and Journey and Construction. I thus drew a figure and framework representing the centrality of journey metaphors in mixed metaphor (as explained below). In this regard, other studies might find another domain that can be a central domain. Thus, they can use this figure and framework, replacing the journey domain with the other domain.

Third, this book can guide us to identify which politicians make climate change a top priority. That is, politicians who predominately use metaphors from certain source domains such as Journey, War and Construction on climate change are more concerned about climate change, where they try to persuade the addressee that it is a serious issue and hence it needs to be addressed than politicians who use metaphors from domains such as Unfairness and Businesses where they try to belittle climate change science, considering commerce and profits the top priority. Also, I have developed a scale indicating the urgency of each type of conceptual metaphor (e.g., war and journey) in comparison to one another.

Finally, all the previous studies of climate change employing CDA investigate the discourse of media coverage, including newspaper, TV programs and reports. However, there has been little attention to politicians' language itself (statements and letters) on climate change, where CDA and Ecolinguistic approaches are employed. I thus analyzed their speeches.

I draw this Fig. 1.1 to show how I used and viewed theoretical terms in this book.

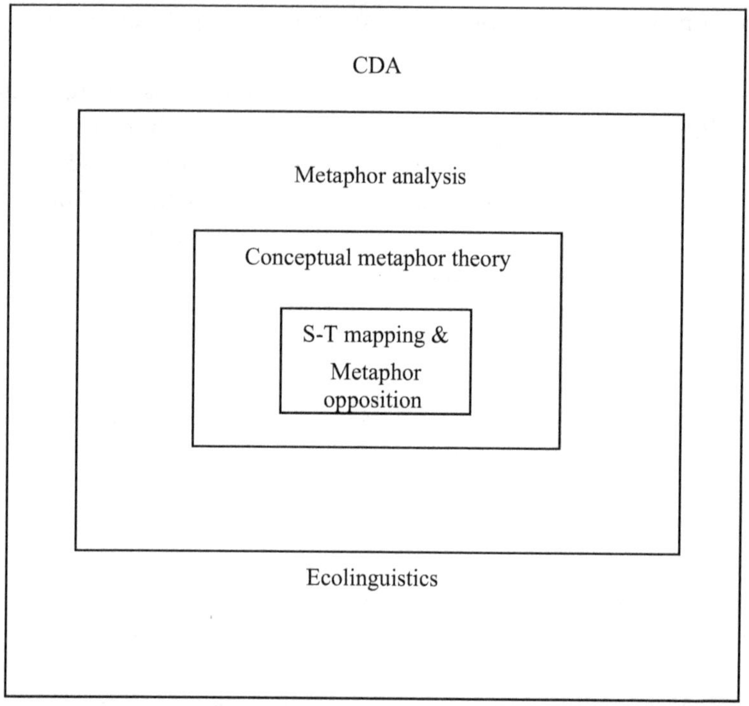

Fig. 1.1 Theoretical terms used in this book

Power is structured as the hierarchy (Fig. 1.2) below indicates. The president, representing the federal level, occupies the top position of power (the federal government), governors and senators, representing the state level (the state government) and mayors representing the local level (the local government) come in the third position. Thus, in the book the social actors are placed in two main hierarchical groups. That is, the federal government versus the state and local government. I shed light on how politicians use language not only to exercise (on the part of the president and those politicians who support the exit from the Paris agreement) but also to challenge power (on the part of the senators, governors and mayors, arguing against this exit) through metaphorical expressions. That is, they exercise power on the audience as they hold authority, and at the same time they resist power exercised by the upper authority (the president or the federal government). In the context of this book, the

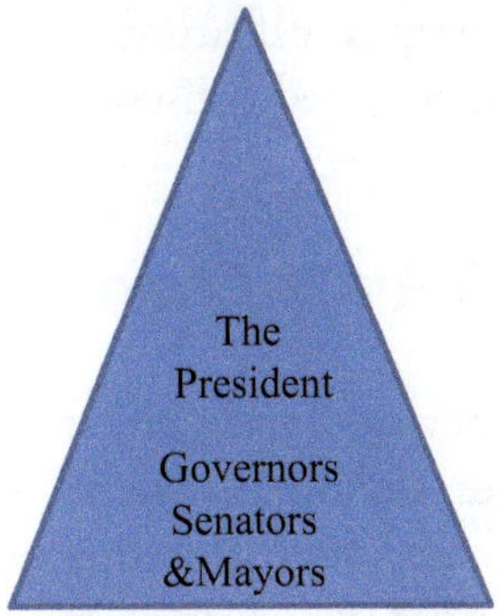

Fig. 1.2 The hierarchy of federal, state and local government

President can access more resources of power than senators, governors and mayors as well as the public. Also, the senators, governors and mayors can access more resources of power than the public. "The most effective use of power occurs when those with power are able to get those who have less power to interpret the world from the former's point of view" (Van Dijk, 1997, p. 184).

As the dangers of climate change threaten our life and health, more and more politicians and journalists are debating and discussing this problem in press conferences, election debates, parliaments and among others. Consequently, climate change is now much more a sociopolitical problem than a scientific problem. Researchers, in turn, in many fields in humanities, including politics, communication, linguistics and social psychology have investigated this problem (including how the problem is discursively constructed and what and how solutions are discursively constructed, and among others) not as a scientific or physical but as a sociopolitical problem, where many realities are created. This book examines how language is used by politicians and in the media to conceptualize climate change (through metaphor). The book demonstrates that metaphor is predominantly used to describe and conceptualize climate change not only by politicians (the president, senators, governors and mayors) but also in the media (the world and U.S. media). Metaphor, as argued by many scholars (e.g., Lakoff & Johnson, 1980), is a conceptual tool that politicians use to create realities, and thereby can influence the addressee's perception of such a problem (climate change). It is, therefore, important to study this linguistic feature in climate change discourse (e.g., in political discourse and media discourse).

An Overview of Climate Change

Climate change is not a new problem, but it has been long debated in politics in many countries around the world, especially in the United States after Trump's announcement to withdraw from the Paris climate agreement. It is a global issue, which is of the utmost concern to most people all over the world. Not only politicians who acted but also the media in the world and U.S. responded to the decision of withdrawal of the U.S. from the Paris climate agreement. Many newspapers, including The New York Times, The Times, The Washington Post and The Guardian devoted a great deal of attention and space to this event (the announcement of withdrawing the U.S. from the Paris climate agreement).

In detail, when the U.S. federal government decided to exit the Paris climate deal, not only did U.S. politicians react but many world leaders, institutions, media and individuals also responded to this decision. Climate change has increasingly become an arguable issue since Trump had been elected as president of the United States.

From the perspective of Ecolinguistics and Critical Discourse Analysis, this book examines language features (metaphors) used by politicians to address and deal with this environmental problem. It examines the ongoing dialogue of negotiation between politicians regarding climate change and the policies affecting it in the U.S. In more detail, on Jun 1, 2017 the American president, D. Trump, announced that he would withdraw the U.S. from the Paris climate agreement on climate change that 195 countries around the world signed. Although the President justified his decision by the claim that the Paris climate agreement will adversely influence U.S. investments and undermine the economy, this decision raised criticisms from many politicians such as senators, governors and mayors who formed alliances against this decision. However, there were many politicians who supported the U.S. withdrawal from the Paris climate agreement. These politicians on both sides used rhetorical devices to influence the public's opinion and policies about climate change in the U.S. The most important function of the environmental discourse "is not to be informative but persuasive" (Alexander, 2009, p. 42). Thus, this book aims at investigating how language is used by politicians to reveal how climate change is conceptualized in the U.S. politics and how climate change is represented in the media.

Thus, the global nature of the phenomenon, which is reflected in the participation of almost all the world's countries in the Paris climate agreement, explains why this issue also takes on a great deal of prominence in the media and why politicians constantly bring this problem into light in their debates, especially over the period between 2015 and 2020. In this regard, any heated issue in the United States may easily attract people's attention from all over the world, including politicians, journalists, and researchers. Reframing Climate Change from Science into Politics, this book presents climate change as a linguistically constructed problem which is represented in the politics of the United States. Politicians have a dramatic impact on policies because they affect our attitudes through their language choices. This is not only true at the federal level, but also at the state and municipal levels. As the problem has been widely debated in social and political domains, many researchers (e.g., Hoffman, 2012) maintain that while scientific consensus on climate change exists, there is no social or political consensus on this phenomenon. This is attributed to the fact that, unlike in science, knowledge is socially constructed in politics, and therefore "what is knowledge for one group may be seen as an ideology by others" (Van Dijk, 1997, p. 28). Hoffman (2012) reports that "surveys show that the American public's belief in the science of climate change has mostly declined over the past five years, with large percentages of the population remaining skeptical of the science" (p. 30). Simply put, while some people advocate for employing more measures to deal with climate change, others try to mitigate the risks or even deny the phenomenon, making the issue seem much less dangerous than it is.

According to Lakoff (2010), the term global warming was replaced by "climate change" in 2003 by Frank Luntz, who was an advisory to the Bush administration. This replacement was justified by the claim that climate change is less frightening than global warming. That is, "the idea was that "climate" had a nice connotation more swaying palm trees and less flooded out coastal cities. "Change" left out any human cause of the change. Climate just changed. No one was to blame" (Lakoff, 2010, p. 71). However, climate change or global warming was originally emerged and debated as a scientific issue, and language used to present this issue was purely scientific. As was the case with many issues such as gun control, abortion, Marijuana and taxes, this problem (climate change) transformed to and resituated in political contexts by politicians in the U.S., especially Republicans and Democrats. In this regard, Carvalho (2018) states that "scientists were the ones who first brought [climate change] into light and will continue to play a key role…., and climate change has been subjected to variegated interpretations in the realm of politics, governmental and otherwise, partly via the media" (p. 488). In the same context, Penz (2018) asserts that climate change is reframed by emphasizing certain aspects of reality and ignoring other aspects through language. For example, politicians and journalists use metaphors from different domains, such as War and Journey, to shape realities that they want their audiences to perceive so that they can legitimize and/or delegitimize normative claims about climate change and its consequences (Fløttum, 2017). Climate change is thus reconstructed as a social and political problem in the language of politicians and the media as well.

When the President officially announced that he would withdraw the U.S. from the Paris climate agreement on June 1, 2017, many politicians in the United States opposed this decision. This book investigates how governors, mayors, and senators reacted to this decision. More specifically, this book investigates how climate change is framed in the language of politicians. In addition, I investigated how climate change is framed in the media in the corpus analysis chapter, and correspondingly I searched for the lexical items (nouns and verbs) that are most collocated with the climate change terms in the U.S. and world media coverage before and after the announcement. I also investigated the frequency of climate change/global warming and Paris agreement terms before and after the announcement.

Paris Climate Agreement

As withdrawing from the Paris climate change serves as the motive for those politicians to issue statements on climate change, this leads us to provide more information about the Paris climate agreement. In this regard, the Paris climate agreement was signed by most, if not all, countries (see Appendix A) around the world for the purpose of addressing climate change (according to the United Nations). This agreement reflects its members' solid and unified stance towards dealing with climate change by calling for more serious procedures and policies that need to be implemented to address this alarming issue. "This includes requirements that all Parties report regularly on their emissions and on their implementation efforts" (From treaties.un.org). And most of, if not all, the world leaders, who signed the Paris climate agreement, seek to minimize its effect at present and in the future. Fløttum (2017) says that "the document itself may be seen as a general action guide for the 195 countries, including a number of procedures and measures that require strict attention, accounting, transparency, and numerous follow-ups" (p. 131). For the date that the agreement entered into force:

> The Paris Agreement was adopted on 12 December 2015 at the twenty-first session of the Conference of the Parties to the United Nations Framework Convention on Climate Change held in Paris from 30 November to 13 December 2015. In accordance with its article 20, the Agreement shall be open for signature at the United Nations Headquarters in New York from 22 April 2016 until 21 April 2017 by States and regional economic integration organizations that are Parties to the United Nations Framework Convention on Climate Change. (treaties.un.org)

The agreement includes 29 articles. These articles serve as codes and regulations for its members. It aims at combating climate change and minimizing its effects. One nighty five countries signed and joined this deal over the period between 2014 and 2016. For the United States, it joined on 22 April 2016. The media often refers to this climate agreement by three terms: Paris agreement, Paris accord and Paris deal. However, the term 'Paris agreement' is more widely used than the other two terms according to the frequency of each of these terms in the NOW corpora.

In early 2017, the President announced that he would withdraw the U.S. from the Paris agreement. On August 2017, the representative of the United States to the United Nations Nikki R. Haley delivered an official notification to the Secretary-General of the UN, stating that the U.S. would withdraw from the agreement (see Appendix B). According to article 28 in this agreement, any country can withdraw from the Paris climate agreement but no earlier than three years after the agreement becomes effective. This means that the U.S. government will remain a member until 2020. "At any time after three years from the date on which this Agreement has entered into force for a Party, that Party may withdraw from this Agreement by giving written notification to the Depositary" (Article 28 from the Paris agreement).

Many political leaders reacted to the decision from both sides. That is, while there were many politicians who urged Trump to withdraw from the Paris agreement and who also supported this decision, there were many other politicians who opposed the decision, but they called for working seriously to deal with climate change.

Political State of Affairs Between 2015 and 2017 in the United States

This section introduces the political structure with regard to the sources of power: Branches of the U.S. Federal Government

"The Federal Government is composed of three distinct branches: legislative, executive, and judicial, whose powers are vested by the U.S. Constitution in the Congress, the President, and the Federal courts, respectively." (whitehouse.gov). The legislative branch (Congress—House of Representatives and Senate) makes laws, the executive branch (president, vice president, Cabinet, most federal agencies) carries out laws and the judicial branch (Supreme Court and other courts) evaluates and interprets laws.

The President has the power either to sign legislation into law or to veto bills enacted by Congress.... and the President has the power to negotiate and sign treaties, which also must be ratified by two-thirds of the Senate. The President can issue executive orders, which direct executive officers or clarify and further existing laws. The President also has unlimited power to extend pardons and clemencies for federal crimes, except in cases of impeachment (whitehouse.gov).

Levels of the U.S. Government

According to the U.S. constitution, there are three main levels of government: Federal, state and local governments. Thus, "powers not granted to the federal government are reserved for states and the people, which are divided between state and local governments" (whitehouse.gov).

Authority is exercised through these levels of the government. "The various levels of government-federal, state, and local-are tied together by a variety of factors: money, programs, political parties, and the play of interest groups among them" (Berman, 2015, p. 20).

For state governments "All state governments are modeled after the federal government and consist of three branches: executive, legislative, and judicial".

Governors, all of whom are popularly elected, serve as the chief executive officers of the fifty states and five commonwealths and territories. As state managers, governors are responsible for implementing state laws and overseeing the operation of the state executive branch. As state leaders, governors advance and pursue new and revised policies and programs using a variety of tools, among them executive orders, executive budgets, and legislative proposals and vetoes (from governors' power& authority).

Thus, they can have a key role in affecting any executive order issued by the federal government. "In every state, the executive branch is headed by a governor who is directly elected by the people." "Governors are the epicenter of politics in the American states" (Ferguson & Foy, 2014, p. 229).

As for Local governments, they work within their state governments "Much of state governments' day-to-day involvement in local affairs occurs through administrative supervision" (Berman, 2015, p. 33). Additionally, "whereas the federal government and state governments share power in countless ways, a local government must be granted power by the state. In general, mayors, city councils, and other governing bodies are directly elected by the people" (obamawhitehouse.archives.gov).

References

Alexander, R. (2009). *Framing Discourse on the Environment: A Critical Discourse Approach*. Routledge.

Berman, D. R. (2015). *Local Government and the States: Autonomy, Politics and Policy: Autonomy, Politics and Policy Authors*. Routledge.

Carvalho, A. (2018). Climate change and the socio-ecological crisis. In A. F. Fill & H. Penz (Eds.), *The Routledge Handbook of Ecolinguistics* (pp. 487–499). New York, NY: Routledge.

Charteris-Black, J. (2018). *Analyzing Political Speeches: Rhetoric, Discourse and Metaphor* (2nd ed.). Palgrave Macmillan.

Fairclough, N. (2014). *Language and Power* (3rd ed.). Routledge.

Ferguson, M., & Foy, J. J. (2014). Gubernatorial Backgrounds, History, Elections, Power. In R. G. Niemi & J. J. Dyck (Eds.), *Guide to States Politics and Policy* (pp. 229–242). Sage.

Fløttum, K. (2017). Willingness of Action. In K. Fløttum (Ed.), *The Role of Language in the Climate Change Debate* (pp. 113–129). Routledge.

Hoffman, A. (2012). Climate Science as Culture War. *Stanford Social Innovation Review, 10*(4), 30–37.

Lakoff, G. (2010). Why It Matters How We Frame the Environment. *Environmental Communication, 4*(1), 70–81.

Lakoff, G., & Johnson, M. (1980). *Metaphors We Live By*. The University of Chicago Press.

Penz, H. (2018). Global Warming Or Climate Change. In A. F. Fill & H. Penz (Eds.), *The Routledge Handbook of Ecolinguistics* (pp. 277–292). Routledge.

State & Local Government. https://obamawhitehouse.archives.gov/1600/state-and-local-government

The White House. Our Government. https://www.whitehouse.gov/about-the-white-house/our-government/

The Withdrawal from the Paris Agreement. (2017). Treaties.un.org/doc/Publication. https://treaties.un.org/doc/Publication/CN/2017/CN.464.2017-Eng.pdf

United Nations. https://unfccc.int/files/essential_background/convention/application/pdf/english_paris_agreement.pdf. The Paris agreement.

USA Gov. Retrieved from https://www.usa.gov/

Van Dijk, T. (1997). *Discourse as Social Interaction*. Sage.

2

Climate Change Discourse/ Politics in Linguistic Studies

The second chapter presents numerous linguistic studies dealing with climate change discourse/politics. It discusses various studies about climate change in terms of how the problem and its aspects are reconstructed. These studies shed light on the role of language in constructing meanings and ideologies about such issues. They draw upon CDA and/or ecolinguistics approaches to study how language is used to frame climate change. In light of these studies, some gaps were identified. This book thus develops metaphor theory and analysis from cognitive perspectives and sheds light on politicians' language itself. This chapter also shows the aims and objectives of this book.

Climate Skepticism

Roper et al. (2016) argued that climate skeptics frequently have used similar communicative (discursive) strategies and tactics against anthropogenic global warming for a long period of time to politicize this issue. Consequently, such tactics have had strong impacts, producing ideologies that can serve their goals and agendas (e.g., delaying any political action to deal with climate change by maintaining a debate so that climate

O. K. Al-Shboul, *The Politics of Climate Change Metaphors in the U.S. Discourse*,
https://doi.org/10.1007/978-3-031-19016-2_2

change is not perceived as a fact but merely as an ideology). Accordingly, the researchers aim to deconstruct these discursive tactics, in an attempt to weaken such impacts through the process of articulation for the sake of disrupting consensus against anthropogenic global warming by discursively constructing the AGW thesis. They echoed Fairclough's (1992) arguments that "the potential danger of this type of critical understanding is that the tactics become available for others" (p. 778). They conducted their analysis to publications, which were 1215 news texts, blogs and reports posted (on the website of the NZCSC from which data was collected) about the New Zealand Climate Science Coalition which, as the researchers claimed, has strong links to the U.S. conservative Think Tanks. They followed principles of critical discourse analysis as put forward by Fairclough (1989, 1992, 2014), paying special attention to linguistic construction, where texts are analyzed in historical and political contexts.

Another important study on climate change is Fleming et al. (2014). They discussed three examples of discourses of climate change, the logical action discourse, the complexity discourse and the culture of consumption discourse. They asserted the role of language in framing the actions that can have powerful impacts, and thereby can be considered. They argue that the key contributor to positive social change with regard to climate change policies is to recognize the power of language in shaping worldviews and then actions and new conditions. And when these worldviews and actions are recognized, discourses can be challenged and reproduced (discursive practice) so that new aspects of knowledge and reality are presented and disseminated. The researchers maintain that in order for the analysis to be effective and crucial, the first step is recognizing the effects of language in particular situations and using the type of language that is more inclusive.

In an attempt to uncover how the U.S. media deals with climate change, Bailey et al. (2014) compared the U.S. coverage on climate change with another country (Spain). They investigated how grammatical and word choice can play a role in constructing uncertainties in four U.S. and Spanish newspapers between 2001 and 2007 for the purpose of determining some articulations linked with uncertainties that portray climate science in the current and future media. They analyzed some language features that reveal such uncertainties. One of the most salient findings concerning the linguistic differences is that the density of epistemic markers in the U.S. newspapers significantly exceeded that in Spanish newspapers, which supports that researchers' prediction that the U.S. press emphasizes uncertainty over the Spanish national newspapers by using a greater number of epistemic markers.

Many studies have drawn on language features as one way to realize how politicians formulate policies to deal with climate change. For example, Fløttum (2010) analyzed two texts drawn from the IPCC Summary on climate change at micro and macro linguistic levels. In this analysis, she examined different linguistic features which were discursively used by the participants. These features are personal pronouns, knowledge, claims, epistemic modifiers, and markers of concession and refusal in polyphonic use. She also investigated the processes at which texts are (re) produced as the participants draw on scientific, public and political discourses. She (Fløttum, 2010) aimed to analyze how social actors at different linguistic levels construct their climate change politics in a discursive manner through language due to the fact that "much political action is, either wholly or partly, linguistic action" (Semino, 2008, p. 85).

Another study conducted on policies of dealing with climate change is by Fløttum and Gjerstad (2012). This book investigates linguistic features at the macro level (narrative) and micro level (sentence/word level) of the South-African Green Paper, which was drafted in 2010 by the Department of Environmental Affairs in South Africa to make a contribution to the world effort to reduce gas emissions. They state their hypothesis that the Green Paper, which includes action plans, is a narrative involving, besides the authors', a number of voices that can be captured and realized through linguistic markers of polyphony, such as negation, sentence connectives, adverbs and reported speech. The researchers argue that by applying the polyphonic (or multi-voiced) theory-the Scandinavian Theory of Linguistic Polyphony, implicit or hidden voices can be revealed, and consequently social actors whose voices are originally hidden (e.g., victims) are brought into light. Hence, this polyphonic perspective has a value in itself.

CDA Analysis into Climate Discourse

Many researchers (e.g., Calliari, 2016; Carvalho & Gupta, 2012; Castro-Sotomayor, 2016; Ebim, 2016; Nerlich et al., 2010; Sedlaczek, 2016; Wills, 2017) applied CDA to their studies on climate change in political and social contexts to study how this issue is identified within the social world. Castro-Sotomayor (2016) maintains that climate change has sparked heated debate in the U.S, making the country more divisive than in any other part of the world, with the result that this phenomenon has gained a prominent position in politics, public communication and media agendas. Such opposing arguments that have been raised about this sensitive issue have led many researchers of different disciplines, especially critical discourse analysts, to analyze the way language is used to address this issue under CDA and Ecolinguistics on the assumption that these approaches examine language use as a form of social practice where meaning, ideologies, worldviews and knowledge are socially constructed by speakers/writers in different discourses. That is, language does not simply reflect reality, but also has a vital role in shaping (social) reality (Fairclough, 2014; Gee, 2014).

Applying ecolinguistic and CDA approaches, Ghdaye et al. (2016) investigated visual texts (advertisements). They examined relationships between media and consumers, and between people and the environment in ecological and political contexts. They revealed how these multimodal texts framed *stories we live by* drawing on linguistic and multimodal features. The researchers aimed at exposing negative uses of language in these texts which promote destructive discourses that destroy rather than conserve nature. They aim to raise people's awareness of how to protect the environment.

Another study which investigates how language used in constructing realties about environment is Ebim (2016). He examined language used on oil explorations and despoliations in Ogonil, from ecolinguisctic and CDA perspective. He applied the theoretical framework of Eco Critical Discourse Analysis to expose the discourses promoted and perspectives constructed in the texts examined. He employed Charles Sanders Peirce's *triadic analytic approach* (icon, an index and a symbol) to study those texts. The data collected is (5) images of events in Ogonil from two Nigerian newspapers. These Newspapers are: The Vanguard and The Punch. The findings indicate that there are many messages loaded in the images, which reveal common thematic preoccupations: neglect, pollution, agitation, environment degradation and despoliation. He argues that this book would call the authorities' attention to what has been damaged with regards to both human and environmental resources. This, as he further argues, enhances collaboration between the authority (the government) and the disadvantaged. He concludes that CDA positively intervenes as it contributes to raising awareness and thereby provides useful resources for those who may be disadvantaged by exposing the hidden ideologies.

Also, Sedlaczek (2016) developed a framework which is in alignment with the Discourse-Historical Approach to critical discourse analysis (DHA) for the purpose of analyzing multimodal representations of climate change in a TV documentary program from the ecolinguistics and critical discourse analysis perspective. The researcher identified six discursive strategies used to represent climate change in the program: argumentation, nomination, predication, perspectivization, and intensification or mitigation. These strategies are discursively used in the genre studied to co-construct knowledge and meaning. In her study, she demonstrates how the scientific knowledge may undergo many processes on the part of media. These processes in turn affect the way knowledge is constructed with the result that reality might be somewhat reshaped. She argues that scientific knowledge especially that is with public interest can be transferred into different but related spheres (like the realms of media, politics and the public). During such a transfer between those realms, it is further subjected to reshaping while discursive struggle over symbolic meanings between social actors and their divergent ideologies, values and interests is going. Therefore, she argues that when media representation of climate change is investigated, it is necessary to consider these conditions and processes of the constructions of knowledge and meanings.

Some researchers combined CDA with corpus methods. Wills (2017), for example, investigated how politicians in the UK constructed and conceptualized climate change in their speeches about Climate Change Bill which was issued in 2008. She conducted corpus analysis combined with critical discourse analysis. For the data of her study, she built two new corpora (CCB and the Budget Corpus), and she used an already existing corpus (the BNC spoken sample). Based on the analysis where she used three corpus methods (keyword analysis, collocation as well semantic groupings) she found that politicians tried to mitigate the consequences of climate change, presenting it as manageable and unthreatening problem. She found that politicians were highly selective of science in ways that serves their ideology that climate change is unthreatening. They also almost avoided discussing people and the environment in their speeches. Additionally, Grundmann and Krishnamurthy (2010) collected news articles about climate change for four countries, namely, US, Canada, UK and France from Nexis over a period between 1984 and 2007. They investigated how the phenomenon gained importance through looking at the distributions of the articles over the period. They analyzed the data using the corpus analysis, from CDA perspective, in terms of the number of articles, word frequency and collocation lists for each one of these countries. Among the main findings is that they found that the number of articles about climate change in French and German print media is about higher five times than in the U.S. and UK. However, the sharp increase in the number of the articles on climate change for the four countries was between 2005 and 2007.

Experimental Research into Climate Discourse

Other researchers also conducted experimental studies to examine people's perceptions of urgency to act on climate change (e.g., Flusberg et al., 2017). They investigated if the public's attitudes towards climate change can be changed due to metaphorical framing (WAR and Race) versus non-metaphorical framing. They found that participants, who were 3000, perceived more urgency and expressed a greater willingness to deal with climate change when they read texts with War metaphors than texts with race metaphors or non-metaphorical texts.

Climate change is a phenomenon which is frequently discussed and debated in many political domains, and therefore language has been extensively studied and examined under many approaches. "The importance of language is further emphasized through… the complexity of the phenomenon of climate change itself, developed from being understood mainly as physical to becoming political, social, cultural, ethical, and communicational" (Fløttum, 2017, p. 1). In the same context, she argues that "since climate change—in contrast to the weather and the consequences of climate change—cannot be seen, heard, or touched, it is in fact through language that we can acquire knowledge of and understand this complex phenomenon" (Fløttum, 2017, p. 7). Concerning this role that language plays in addressing environmental issues, Penz (2018) asserts that it works in multiple ways. That is, although speakers might use language in a way that looks neutral, they might hide some environmental exploitation or for example they might describe a phenomenon in a more negative way than the reality (than it really looks like). In fact, many researchers emphasize the role of language in addressing many environment-related problems. For example, Fill and Penz (2018) assert that "language is crucial in creating an awareness of environmental problems and of the processes which lead to climate change" (p. 438). They report that there is an increasing interest in ecolinguistics research in the U.S. and many countries around the world, especially with the growing demand for a political and public action to confront this environmental phenomenon, after it was almost restricted to North America and Europe. One important reason why CDA and ecolinguistic approaches are important in investigating climate change policies is that organizations manipulate language to create agendas to reject and marginalize those who call for change (serious measures to combat climate change to suppress and discredit the scientific consensus on *anthropogenic climate change* in the United States, as argued by Alexander, 2018). Thus, climate change is not only a scientific problem but also a social problem where the role of collective identity should be emphasized for the purpose of eliciting reactions and triggering social change (Priest, 2016). In this respect, Fløttum (2017) asserts that making the transition from *a high carbon-society* to *a low-carbon society*, in an attempt to combat climate change, is not only the responsibility of decision-makers (e.g., governments) but also citizens. That is, their attitudes towards such transition are important since these attitudes, which may be acquired and adopted through *public discourse of politicians*, can become, and lead to, action, and thereby can influence policy makers by pushing them to take specific procedures to deal with climate change. This is all performed by language. The most important aim of politicians (e.g., the governors and mayors), is to get people to hold views and these views may transform to actions "i.e. by getting others to hold views (that may lead to action) that are advantageous to a particular individual, group or cause" (Semino, 2008, p. 85). This demonstrates why language is important. In fact, many researchers (e.g., Fløttum, 2017) throws light on the role of language. They call for producing language that can be easily understood to ordinary people so that they can become active participants in decision-making processes on different environment-related problems. In this regard, Nerlich et al. (2010) investigated the role of language in climate change communication. They assert that this role cannot be neglected when climate change communication is investigated. They (Nerlich et al., 2010) reported that scientific communicators are constantly encouraged by commentators, especially politicians to adapt their language to that of ordinary people to make it easier to be understood by the public, and thereby people are more able to take a stance on this environmental issue so that they can become active participants. The article generally highlights the role of language (including particular metaphors, frames, and narratives) in conveying climate change issues to stakeholders. In this regard, it is important to point that there are some agendas why politicians simplify language used on climate change. They want people to have an active role in triggering social change to confront climate change (e.g., calling for employing more serious policies to fight climate change), and thereby politicians can be supported by the public so that they can strengthen their arguments.

This book analyzed language features (metaphors) as conceptual tools used by politicians to reframe this environmental issue (climate change) as a sociopolitical problem. They used metaphors drawn from physical and everyday experiences of life as the addressee is familiar with such physical experiences. This in turn facilitates people's perceptions of realities shaped by politicians (through using a metaphor) as these realities can be understood in terms of concrete domains such as journey. This book further investigates how speakers perceive realities and act accordingly through two main cognitive schemas (mapping and opposition) from a cognitive perspective. This book thus enriches linguistic analysis with cognitive aspects to that I can offer deeper explanations for metaphor analysis.

Aims and Objectives

This book aims to enrich linguistic analysis, guided mainly by CDA and Ecolinguistics, with cognitive aspects of analysis as one way to offer deeper explanations of how politicians create realities about climate change (how climate change is represented?) and how these realities can be perceived by the addressee, and how he/she can act accordingly. This book, therefore, draws on cognitive analysis (particularly source-target domain mapping and metaphor opposition), offering a better understanding of critical metaphors. This book investigates how language (metaphor) can shape realities for the sake of triggering social change. In this regard, Fairclough (2014) argues that "metaphor is not merely a way of viewing realities; it constituted a license for policy change and political and economic action" (p. 156). This book also examines how these realties can be perceived by the addressee.

Many previous studies of climate change employing CDA investigate discourse of media coverage, including newspaper, TV programs and reports. However, there has been little attention to politicians' language itself on climate change, where CDA and Ecolinguistic approach are employed (a detailed analysis of politicians' discourses). Accordingly, this book examines metaphors employed by politicians themselves: the president, the senators, governors and mayors. It aims to find out what metaphorical concepts that can be realized from the metaphorical expressions used by the social actors (the president, senators, governors and mayors). Thus, in this book I want to fill that gap and examine how politicians frame climate change and in so doing create 'realities' for their audience. I discussed these concepts at the interpretation stage to show how climate change is conceptualized in the U.S. politics and what entailments that they imply so that I can explain (what) social changes that these metaphorical expressions aim to trigger at the macro level. Furthermore, there is much research investigating metaphor in social and political texts and there are few studies which conduct linguistic analysis into metaphor in climate change discourse. This book, therefore, goes deeper by employing CDA and ecolinguistics to investigate how climate change is represented in the U.S. politics. It aims to offer deeper explanations of how realities can be shaped not only from source-target mappings but also using the metaphor opposition. This cognitive approach (the metaphor opposition) aims at further explaining how realities can motivate the addressee to act according to such realities through introducing favored (positive) and unfavored (negative) conceptual entailments that can be inferred from the metaphors used as a cognitive framework for the sake of explaining how metaphor triggers people. These elements can direct the addressee to act according to these realities.

These book questions investigate how climate change is framed. These questions include how climate change is represented in the U.S. politics through metaphor? (How are metaphors employed by politicians addressing climate change?) What are the metaphorical expressions and the conceptual metaphors of climate change (that can be realized by the metaphorical expressions) in the data? What are the most-and least-common conceptual metaphors? And why? Which conceptual metaphors stress a stronger urgency to respond to climate change than the others? What are the sociocultural views (e.g., the notions that are foregrounded as realities to motivate and persuade people to support climate change policies in the U.S?) formed? The questions of the corpus analysis chapter provide statistical evidence of how climate change is conveyed before and after the announcement of withdrawing the U.S. from the Paris climate agreement in the world and U.S. media through investigating the two following main questions. How is climate change represented in the world and U.S. media? What kinds of aspects and actions that the world and U.S. media focuses on with regards to climate change? Did the President's decision of withdrawing from the Paris agreement spark heated debate? (did the climate change/global warming and Paris agreement/accord/deal terms increased in the world and U.S. media after the president's announcement to withdraw from the Paris climate agreement? Are the differences statistically significant? Accordingly, these all questions stated above investigate the language of politicians about climate change. They also examine the media texts on climate change using corpus analysis in a separate chapter.

References

Alexander, R. (2018). Investigating texts about environmental degradation using critical discourse analysis and corpus linguistic technique. In A. F. Fill & H. Penz (Eds.), *The Routledge Handbook of Linguistics* (pp. 196–210). New York, NY: Routledge.

Bailey, A., Giangola, L., & Boykoff, M. (2014). How Grammatical Choice Shapes Media Representations of Climate (Un)certainty. *Environmental Communication, 8*(2), 197–215.

Calliari, E. (2016). Loss and Damage: A Critical Discourse Analysis of Parties' Positions in Climate Change Negotiations. *Journal of Risk Research, 21*(6), 1–23.

Carvalho, A., & Gupta, J. (2012). State Commitment to Promoting Public Participation: The UNFCCC and Citizens' Roles in Climate Change Politics. In A. Carvalho & T. R. Peterson (Eds.), *Climate Change Politics: Communication and Public Engagement* (pp. 123–161). Cambria Press.

Castro-Sotomayor, J. (2016). Communicating Climate Change: The Path Forward. Book Review. *Language & Ecology, 1–3.*

Ebim, A. M. (2016). An Ecolinguistic Study of Media Deception of Oil Spillage in Ogoniland. *Language & Ecology*, 1–19.

Fairclough, N. (1989). *Language and Power* (1st ed.). Pearson Education Limited.

Fairclough, N. (1992). Discourse and Text: Linguistic and Intertextual Analysis Within Discourse Analysis. *Discourse & Society, 3*(2), 193–217.

Fairclough, N. (2014). *Language and Power* (3rd ed.). Routledge.

Fleming, A., Vanclay, F., Hiller, C., & Wilson, S. (2014). Challenging Conflicting Discourses of Climate Change. *Climate Change, 127*(3), 407–418.

Fløttum, K. (2010). Linguistic and Discursive View on Climate Change Discourse. *Groupe D'étude Et De recherche Enanglais De Spécialité, 58*, 19–37.

Fløttum, K. (2017). Willingness of Action. In K. Fløttum (Ed.), *The Role of Language in the Climate Change Debate* (pp. 113–129). Routledge.

Fløttum, K., & Gjerstad, O. (2012). Arguing for Climate Policy Through the Linguistic Construction of Narratives and Voices: The Case of the South-African Green Paper "National Climate Change Response". *Climatic Change, 118*, 417–430.

Flusberg, S., Matlock, T., & Thibodeau, P. (2017). Metaphors for the war (or race) against climate change. *Environmental Communication, 6*, 769–783.

Gee, J. P. (2014). *An Introduction to Discourse Analysis: Theory and Method.* Routledge.

Ghdaye, K., Bailo, V., Fortunato, E., Gadient, S., Gosgnach, I., Mantellato, M., Napolitano, M., Rinaldi, F., & Roman, F. (2016). Ecolinguistic Perspectives on Our Surroundings. *Language & Ecology*, 1–20.

Grundmann, R., & Krishnamurthy, R. (2010). The Discourse of Climate Change: A Corpus-Based Approach. *Critical Approaches to Discourse Analysis Across Disciplines, 4*(2), 125–146.

Nerlich, B., Koteyko, N., & Brown, B. (2010). Theory and Language of Climate Change Communication. *Climate Change, 1*, 97–110.

Penz, H. (2018). Global Warming Or Climate Change. In A. F. Fill & H. Penz (Eds.), *The Routledge Handbook of Ecolinguistics* (pp. 277–292). Routledge.

Priest, S. (2016). *Communicating climate change: The path forward*. UK: Palgrave Studies in Media and Environmental Communication.

Roper, J., Ganesh, S., & Zorn, T. (2016). Doubt, Delay, and Discourse: Skeptics' Strategies to Politicize Climate Change. *Science Communication, 38*(6), 776–799.

Sedlaczek, A. (2016). Representations of Climate Change in Documentary Television: Integrating an Ecolinguistic and Ecosemiotic Perspective into a Multimodal Critical Discourse Analysis. *Language & Ecology*, 1–19.

Semino, E. (2008). *Metaphor in Discourse*. Cambridge University Press.

U.S. Census Bureau. https://www2.census.gov/geo/pdfs/maps-data/maps/reference/us_regdiv.pdf

United Nations. https://unfccc.int/files/essential_background/convention/application/pdf/english_paris_ag reement.pdf. The Paris agreement.

United States Climate Alliance. https://www.usclimatealliance.org/alliance-principles

USA Gov. https://www.usa.gov/

Wills, R. (2017). Taming the Climate? Corpus Analysis of Politicians' Speech on Climate Change. *Environmental Politics, 26*(2), 212–231. https://doi.org/10.1080/09644016.2016.1274504

3

A Detailed Overview of Theory: CDA, Ecolinguistics and Metaphor

The third chapter presents theoretical foundations of the main theories used in this book, including CDA, Ecolinguistics, and Metaphor. These concepts are presented and explained to constitute an understanding of how the speaker makes meaning (e.g., realities about climate change) and how these realities can be perceived by the audience. This chapter also discusses the relationship between CDA and ecolinguistics and how metaphor analysis can be based on those two theories. This chapter also presents the main concepts of these theories. Those concepts and frameworks from these theories are important to establish a solid background for the data analysis.

The Theoretical Foundations: Critical Discourse Analysis

The third chapter presents the theoretical foundations of CDA, Ecolinguistics, Metaphor. These concepts are discussed theoretically to constitute an understanding of how the speaker makes realities and these realities can be perceived by the audience.

O. K. Al-Shboul, *The Politics of Climate Change Metaphors in the U.S. Discourse*,
https://doi.org/10.1007/978-3-031-19016-2_3

CDA is among the most prominent theories that are employed in discourse studies. This is the clearest way to examine the relationship between political language choices and social realities. Van Dijk (1993) maintains that "critical discourse analysts want to know what structures, strategies or other properties of text, talk, verbal interaction or communicative events play a role in these modes of reproduction" (p. 283) such as political inequality and power abuse. Rogers (2011) and Wodak and Meyer (2008) maintain that critical discourse analysis originally emerged and developed in critical linguistics guided first by principles from this field particularly in Language and Control by Flower et al. (1979) and Language and Ideology. The term 'critical' was drawn from critical linguistics; and 'discourse analysis' was used instead of 'linguistics' (Wodak & Meyer, 2008).

The main development of critical discourse analysis as theory is attributed mainly to Fairclough, 1989, 2001, 2014; Gee, 1999 & 2014; Van Dijk, 2008, 2009 & 2014; Wodak, 2008 who focused their research on the relationship between texts themselves on the one hand and texts and power on the other hand, assuming that there is a dialectical relationship between semiotic features and social elements; or between texts and (social) events.

According to many scholars, CDA is an interdisciplinary approach. In this regard, Wodak and Meyer (2008) maintain that "the manifold roots of CDA lie in Rhetoric, Text linguistics, Anthropology, Philosophy, Socio-Psychology, Cognitive Science, Literary Studies and Sociolinguistics, as well as in Applied Linguistics and Pragmatics" (p. 1). They further say that:

> The CDA as a network of scholars emerged in the early 1990s, following a small symposium in Amsterdam, in January 1991. Through the support of the University of Amsterdam, Van Dijk, Norman Fairclough, Gunther Kress, Theo van Leeuwen and Ruth Wodak spent two days together, and had the wonderful opportunity to discuss theories and methods of Discourse Analysis, specifically CDA (Wodak & Meyer, 2008, p. 3).

Reisigl (2017) argues that the term CDA was used in the late 1980s for the purpose of making a difference between descriptive discourse analysis and CDA. CDA theory, however, is now applied to study many current issues especially in political contexts, like climate change. According to Fairclough (2014) and Wodak and Meyer (2016), CDA can only be applied to social and political problems since it investigates the role of discourse in the way power is used and in the (re)production of hegemony. CDA is a multidisciplinary approach which aims at paying special attention to the role of discourse (e.g., language features and the discourse structure) in the construction of the social world (e.g., ideologies, knowledge, power abuse, etc.), as argued by many CDA scholars (e.g., Fairclough, 2014; Gee, 2014; Van Dijk, 1995; Wodak & Meyer, 2008). Furthermore, CDA has witnessed many developments, with the result that it has been applied to conduct a more detailed linguistic analysis for the purpose of revealing the role of language in creating realities and myths about social issues such as climate change, abortion and gun control. According to Hart (2018), among the most prominent developments are these that pertain to critical applications of cognitive linguistics, where it (cognitive linguistics) investigates the cognitive aspects or features that texts have. This in turn helps the researcher offer deeper explanations of how meanings are constructed. For example, Charteris-Black's *critical metaphor analysis,* which is CDA-based, pays special attention to the role of metaphors (as ideologically loaded linguistic features) in the construction of social realities from the point of view of how they give us some specific ways of thinking of the social world.

With these developments associated with CDA, it has become easier to draw distinctions between CDA and other approaches to DA, it is argued that approaches to discourse analysis (DA) analyze language (including patterns and features) in use, collecting data from the real world, to see how people interact and how they draw on language features to achieve such goals as enacting and/ reconstructing identity, among others. Thus, this type of analysis more focuses on language in terms of social relationships and communication and how meanings are (re)constructed. Amossy (2018), for example, argues that "DA approaches aim at analytical understanding rather than assessment" (p. 266). But, CDA critically investigates how language is used to enact power. It pays special attention to the way power is abused to produce ideologies and exercise domination. "CDA focuses on the strategies of manipulation, legitimation and other discursive ways to influence the minds of people (and indirectly the actions) of people in the interest of the powerful" (Van Dijk, 1995, p. 18).

Other scholars (e.g., Gee, 1999, 2014; Johnstone, 2008), however, argue that all discourse analysts should be critical, because language itself is political. "My view is that all discourse analysis needs to be critical, not because discourse analysts are or need to be political, but because language itself is political" (Gee, 2014, p. 9). Additionally, Johnstone (2008) investigates power, and the way power is enacted in discourse analysis, looking closely at the relationship between language and the world and how they construct each other. Power is, therefore, an important concept in DA approaches. In this regard, Carvalho (2018) maintains that the American branch of DA emphasizes that discourse is shaped by the world, and discourse shapes the world, and thereby they pay special attention to the notion of power and hegemony in DA studies. "Functionally, discourse is used (simultaneously) to represent, evaluate, argue for and against, and ultimately to legitimate or delegitimate social actions" (Hart & Cap, 2014, p. 1). They (Hart & Cap, 2014) further argue that "there are long-standing traditions in discourse analysis, e.g., post-structuralist discourse analysis, which adopt a critical perspective (e.g., Slembrouck, 2001)" (p. 2). However, one way to distinguish between the two approaches (DA and CDA) is to look more closely at the role of the critical discourse analyst. In this respect, Gee (2014) highlights some important and specific goals for critical discourse analysts:

> Their goal is not just to describe how language works or even to offer deep explanations, though they do want to do this. They also want to speak to and, perhaps, intervene in, institutional, social, or political issues and controversies in the world. They want to apply their work to the world in some fashion (Gee, 2014, p. 9).

By way of explanation, according to many scholars (e.g., Gee, 2014; Van Dijk, 2015), CDA adopts critical perspectives that may be found in other fields of discourse analysis. Nevertheless, unlike DA, CDA demands an overt commitment on the part of the CDA analyst, indicating which side (group) he/she supports (Carvalho, 2018; Fairclough & Wodak, 1997; Tenorio, 2011). That is, "critical discourse analysts take an explicit position and thus want to understand, expose, and ultimately challenge social inequality" (Van Dijk, 2015, p. 466). This means that the critical discourse analysist's attitude can influence critical discourse-based analyses. In this regard, Van Dijk (2015) maintains that "CDA is discourse study with an attitude" (p. 466) since CDA requires "the self-reflection of the researchers themselves" (Fairclough & Wodak, 1997, p. 279). Accordingly, CDA provides explanations and critique on the part of the researcher (Gee, 2014; Sum & Jessop, 2013). This means that the researcher's attitude determines the explicit position that he/ she takes. As this book investigates texts on environmental issues from CDA perspective, this analysis should be based on ecolinguistics, which is concerned with environmental discourse. Thus, this leads us to introduce the main themes in this field (ecolinguistics). The term 'Ecolinguistics' comprises two items: eco and linguistics. Muhlhausler (2018) maintains that the term ecology was first used in linguistics in Voegelin et al's work in 1967: "The first use of the term 'ecology' in linguistics is found in a paper by the Voegelins (Voegelin et al., 1967) on the language varieties in Arizona where a distinction between intra-language and inter-language ecology is drawn" (Muhlhausler, 2018, p. 136). Fill and Muhlhausler (2001) assert that the American linguist Einar Haugen first borrowed the ecology term in 1970, which had been coined by Haeckel (1866) in biology to refer to the relationship between organisms, to refer to the relationship between organisms and language. Since then many scholars reflected this view (of that relationship) in their research. Halliday (2001), for example, argued that "classism, growthism, destruction of species, pollution and the like - are not just problems for the biologists and physicists. They are problems for the applied linguistic community as well" (p. 199). However, "Ecolinguistics is still a very young research area. In 2017, it celebrates its 45th birthday- if we regard Einar Haugen's article in 1972 as its beginning" (Fill & Penz, 2018, p. 437). They (Fill & Penz, 2018) further maintain that this science is considered to be "based on American and European ideologies, especially in the first few decades of its existence" (p. 437). Fill (2018) defines Ecolinguistics as "a kind of ideology that creates an awareness of the interdependence of all things and ideas" (p. 5). As ecolinguistic studies take discourse as the main focus of analysis, it is important to highlight that relationship. In this regard, Stibbe (2014 &

2017) maintains that ecolinguistics is a form of CDA with a difference that ecolinguistics focuses on the larger ecological system to include the relationship of humans with the environment. It reveals the positive and negative roles of humans in the natural world. He further maintains that ecolinguistics distinguishes between two types of discourses: those that protect and those that destroy the environment. In this regard, Stibbe (2017) asserts that Ecolinguistics has an important role in exposing the dominant discourses that "promote ecologically destructive behavior" (p.165); and, in turn, it promotes those that "encourage relationships of respect and care for the natural world" (Stibbe, 2014, p. 117).

Focuses of CDA. In this section, I will highlight seven concepts which influence and make up CDA assessment. They are frequently used in many CDA scholars' works (Charteris-Black, 2014, 2018; Fairclough, 2001, 2014; Gee, 1999, 2014; Van Dijk, 1997, 2003; Wodak, 2014). In this section, I will draw links between these seven concepts so they can be used as a theoretical background in CDA based studies. These concepts are:

- Intertextuality
- Social struggle
- Persuasion
- Power
- Reality
- Legitimization
- Social change

Fairclough (2014) argues that "texts always exist in intertextual relations with other texts, it is arguable that they are always dialogic, a property which is sometimes referred to intertextuality" (p. 166). This, to a large extent, explains why "texts can bear intertextual traces of other texts in many ways, ranging from the most direct repetition to the most indirect allusion" (Johnstone, 2008, p. 164.). This means that the historicity of texts is emphasized in CDA studies as it helps explore a network of texts that focus on one aspect of reality rather than another (Fairclough, 1992; Musloff, 2016; Van Dijk, 2008; Wodak, 2001).

Accordingly, CDA scholars (e.g., Fairclough, 2014) assert that texts are not analyzed in isolation from other texts due to the fact that texts are linked to each other through strategies such as explicit references, evocation, etc. Thus, researchers who employ CDA theory follow an in-depth approach to find and trace how these texts are interrelated. Wodak and Meyer (2016), for example, argue that the discursive differences arising between social actors (e.g., politicians) are negotiated in texts, and consequently texts are considered sites of struggle through which opposing discourses and ideologies can be traced and examined. This brings up the second notion in CDA, which is social struggle. The term 'argumentation' is also used by many scholars and researchers (Reisigl, 2017) as an alternative term for social struggle. Reisigl (2017) maintains that argumentation is an important discursive strategy. The analyst looks at texts, revealing how discourse develops around social problems and reflects the struggle. The social struggle or argumentation around climate change between groups is a good example of this kind of struggle or argumentation where they (groups) compete against each other for the purpose of gaining some political benefits (political exploitations). In this respect, Fairclough (2014) maintains that persuasion and manipulation are among the main goals of argumentation. This explains why "… arguments are ideally designed as specific responses to specific, situated problems. From this perspective, argument is not a kind of conversation, but rather a process that arises in conversation" (Johnstone, 2008, p. 110). Accordingly, argumentation is an important means to persuasion (e.g., to get the addressee adopt the writer/ sparker's attitude). In this regard, persuasion, the third component of CDA, represents the goal that the speaker/the writer tries to achieve. It is an important concept associated with language when it is investigated under CDA. Johnstone maintains that "people use discourse to persuade, to cause others to act, to change the world" (2008, p. 265). Similarly, Semino (2008) maintains that persuasion is an important notion that speakers/writers aim to achieve through language. Charteris-Black (2014) maintains that "various linguistic means can contribute to how convincing, and how convinced, a speaker sounds" (p. 99).

But in order for the speaker/writer to persuade the addressee, the speaker/writer should enjoy some power for that goal. This leads us to discuss the fourth focus, which is power. "Different types of power may be distinguished according to the various resources employed to exercise such power" (Van Dijk, 2015, p. 469). In this respect, Van Dijk (2015) distinguishes between three types of power:

> The coercive power of the military and other violent people will "[…]" be based on force; the rich will have power because of their money; the more or less "persuasive power" of parents, professors, or journalists may be based on knowledge, information, or authority (p. 469).

The last type of power is the one that draws on persuasion. It is the most widespread and common at present. In this type of power, there are two important resources for power: The first resource is the ability to have active access to the public genres such as press conferences, news, interviews or TV talks (Van Dijk, 1993, 2015). This in fact supports Wodak's argument that "language is not powerful on its own – it gains power by the use powerful people make of it" (2002, p. 10). Power, therefore, is derived from those who produce that language. This means that the same utterance can have different impacts (perlocutionary force) on the audience depending on which (social actors) produce that utterance. Thus, one of the researcher's jobs is to investigate the speaker's background like what social groups he/she belongs to and what social positions he/she occupies in that group since this can determine to what extent his/her language is powerful. Van Dijk defines power as a form of control that powerful people can exercise. And one way to exercise such control is via access to specific forms of discourse. Texts, therefore, investigated from CDA perspectives are always assigned to genres; e.g., climate change discourse, which is realized through a range of genres like TV debates that shed lights on the politics of governments, speeches or lectures by experts, and guidelines to reduce energy consumption to study by whom these texts are produced and which have more active access to such resources (Wodak, 2001). Accordingly, CDA is the most effective approach to study media and political discourse as it focuses on power and dominance in particular and reveals the role of discourse in producing such notions. The second crucial resource for power is alliances. In this type, Fairclough (1992) maintains that "the dominant groups also appear to exercise power through constituting alliances, integrating rather than merely dominating subordinate groups and winning their consent" (p. 94). For example, while modern governments avoid using coercion, they resort to persuasion and attraction through language as an effective strategy to convince people of what they need to do, so that they can act in ways that governments or parties desire (Anastassov, 2018; Mulderrig, 2011). This means that politicians exercise power through such resources to get the recipient to believe what they claim, and consequently follow what is called for by them. This brings up the fifth focus of CDA which is reality. CDA theory focuses on *reality* as one of its basic components due to the belief made by many scholars (e.g., Fairclough, 1989, 2014; Gee, 1999, 2014; Lakoff & Johnson, 1980; Van Dijk, 1994) that we use language not only to transmit existing knowledge but also to create new knowledge. Language is, therefore, viewed as a tool to shape realities that speakers/writers want their audience to perceive. For example, "we define our reality in terms of metaphors and then proceed to act on the basis of the metaphors" (Lakoff & Johnson, 1980, p. 158). With regards to reality and its influence on the addressee, many CDA scholars, especially those in the sociocognitive approach (SCA), identify the relationship between discourse structures and mental processes. Van Dijk (2003), for instance, maintains that some specific linguistic discursive features like metaphors might have negative effects on the audience. That is, people might favor one opinion over the other simply due to the speaker's/ writer's language (what and how linguistic features are used in a particular context) with the result that one group's ideology may be adopted as a fact or reality in the interest of that group.

With regards to the type of reality or knowledge, Van Dijk (2003) explains several types of knowledge: declarative; procedural knowledge; personal; general; specific, and social knowledge. However, the type of knowledge that discourse analysis is most concerned with, and thereby the current study, is (social) shared knowledge. This kind of knowledge (e.g., realities made by politicians about climate change) is socially and culturally constructed by the writer/speaker. Thus, knowledge keeps changing and shaping in accordance with the writer/speaker's agendas and goals with the result that we can find many realities created by politicians about an issue despite the existence of a concrete piece of evidence that supports one reality. One of the important characteristics of this type is that knowledge is contextual. This means that what is considered as justified belief in one context may not be so in another context. Simply put, what is regarded as a reality by one group or institution may not be so by another group or institution. This is attributed to the fact that speakers/writers employ linguistic features, particularly metaphor, as an effective strategy or technique for (de)legitimizing myths or realities in political speech (Charteris-Black, 2005; Fairclough, 2014; Lakoff & Johnson, 1980). This takes us to shed light on the sixth focus, which is legitimization. In politics, discourse of legitimization is of high importance because when a politician creates an ideology (e.g., an attitude that he/she needs his/her audience to adopts), he/she needs to legitimize that ideology. Politicians create myths or ideologies through a discourse of legitimization (Charteris-Black, 2005, 2011). And in turn, they delegitimize the other (oppositional) ideology or attitude, as he/she legitimizes his/her own. In this regard, Leeuwen (2018) argues that "de-legitimation plays a crucial role in critiquing oppositional values" (p. 220). This is a necessary discursive practice that politicians employ to make the ideology that they create true, and thereby should be adopted. According to Van Dijk, 1997, what is considered as an ideology for one audience, it can be seen as knowledge to other audience. For this reason, knowledge is social. This means, according to him, what is perceived as knowledge by one group of people, it is not necessary that it is accepted as knowledge by another group.

However, this leads us to ask why politicians aim to shape realities and to present their opinions as realties to the addressee? This takes us to discuss the last (seventh) focus, which is social change. Because CDA theory is a valuable resource to interpret how discourse is, and can be, related to such social elements as power and ideologies, CDA provides explanations of why speakers/writers aim to shape realities. Fairclough (2014) points out that the main goal of shaping realities is to trigger social change through (language). He argues that discourse can lead to political actions "by increasing understanding of existing reality and its problems and possibilities" (Fairclough, 2014, p. 13). Fløttum (2017) gives an important example of social change that some politicians and activists try to make when they address climate change, which is *willingness of action*. That is, such politicians and activists urge actions, through making the addressee perceive their points of view or attitudes as realities, so that they can affect policies on climate change, and thereby triggering social change.

CDA theory, based on such principles, is among the most prominent theories employed to investigate language (e.g., linguistic features) used by speakers/writers like politicians to convince the audience of their arguments, with the result that it has been widely applied to studying language used to address environmental problems to expose ideologies and myths (knowledge) constructed about the issue. By employing CDA theory, I will be able to investigate the relationship between texts and sociocultural meanings, showing how such textual analysis contributes to the social formation of speakers' (social actors') ideologies with regard to climate change. In this book, I also took an ecolinguistic approach, which is CDA-based, as the book is concerned with climate change. When Ecolinguistics emerged, it aimed to study texts about nature and has since developed to look for positive uses of language that would help protect the environment from deterioration and damage (Fill, 2018; Fill & Penz, 2018).

In this book I draw on CDA as theory to investigate climate change in social and political context based on the following general principles and guidelines, formulated mainly by many CDA scholars (e.g., Fairclough, 1989, 2014; Van Dijk, 1993; Wodak & Meyer, 2016) that underline this theory (CDA) and that are in consonance with the assumptions of this book:

1. There is a social problem that needs to be addressed. Climate change is not only a scientific problem but also a social problem. It, therefore, requires *a social movement solution.* As it has developed from being originally scientific to political and social public issue. As a consequence, society should be fully aware of its risks (Fløttum, 2017; Priest, 2016). Accordingly, different discourses produced by the proponents and opponents of climate change policies are analyzed.
2. Language is a social practice. It can create ideologies and realities. This means that politicians try not only to reflect but also to (re)shape realities, and one of the aims of this book is to reveal realities shaped their language (statements and letters). That is, politicians try to co-

construct meaning and knowledge. Therefore, illocutionary rather than locutionary forces of utterances need to be realized. Accordingly, CDA is employed since although it extensively draws on the linguistic analysis (the micro level), it goes beyond the boundaries of linguistic analysis and description to analyze texts in a wider sociopolitical context, analyzing language as constitutive of the social world where the problem investigated, and solutions suggested are socially constructed.

3. Language is used not only to exercise but also to challenge and resist power. The political speeches in this book are produced to exercise power and are produced in turn to challenge power, taking a stance against the federal government's decision, for the purpose of advocating climate change policies using linguistic features that are ideologically loaded.

Approaches to CDA. CDA is a multidisciplinary theory which researchers draw on to analyze discourse. Within the CDA theory, there are approaches outlined by CDA scholars. These approaches serve as guidelines for researchers employing CDA theory. Wodak and Meyer (2016) maintain that CDA is a set of approaches for discourse analysis. However, while these approaches vary in their orientations, they overlap, making up CDA theory (Wodak & Meyer, 2016).

Sociopolitical approach. This approach investigates texts in social and political contexts (Fairclough, 2001, 2014). In this approach, Fairclough emphasizes how texts can constitute a policy for change. Although he focuses on language at micro level (conducting detailed linguistic analysis of texts), he investigates texts in a larger social structure where his approach sheds light on how discourse shapes the social world and how is shaped by the world. His approach, consisting of three analytical levels (micro, meso, ands macro), constitutes an important resource to provide deep explanations at macro level based on linguistic analysis of language.

Sociocognitive approach. Van Dijk (2008, 2009, 2014) emphasizes how language affects our attitudes, and thereby behavior. In this approach, CDA does not limit itself to engaging "in social and political analysis of the context of text and talk" (Van Dijk, 2018, p. 27). But he maintains that "it [CDA] explains how the symbolic elites of politics and the mass media are able to control public discourse and attitudes" (Van Dijk,

2018, p. 27). Accordingly, this approach draws on cognitive aspects of discourse to further explain how attitudes are affected. He investigates the most common and important linguistic features that can be ideologically loaded such as metaphor. In this approach, language is viewed as an important element that plays an important role in affecting and determining how the social world is perceived. In this regard, this approach is more interested in social and political issues that are of public interest.

Discourse-historical approach. There are many principles that are shared by the previous CDA approaches and DHA (Reisigl & Wodak, 2008). But what distinguishes this approach from the other approaches is that it focuses on the historical context (the historical dimensions of discourse formation). That is, the historicity of texts (since texts do not exist in a vacuum) plays a key role in their interpretation. Another important principle in the DHA approach is that "Numerous genres and public spaces as well as intertextual and interdiscursive relationships are studied" (Wodak, 2014, p. 15). The DHA scholars consider context is an integral part of the approach since it facilitates empirical observations and help the analysist develop theories and methods. They altogether form triangulation (observation, methods and theories), which is one of the most distinctive DHA properties. "[DHA] integrates and triangulates knowledge about historical, intertextual sources and the background of the social and political fields within which discursive strategies employed" (Wodak, 2014, p. 529). Under this approach many issues have been investigated, especially in politics like discrimination and racism as well as ecological problem (climate change).

Critical discourse analysis as a theoretical framework involves "detailed examination of texts to reveal hidden ideologies that are subtly conveyed by the use of particular linguistic features" (Fairclough, 2014, p.174). Since this book focuses on the linguistic level (micro level), I adopted this theory for the sake of explaining how several types of metaphors (Journey, War, Cleanliness and Construction) that are ideologically loaded can lead to the social construction of the problem and solutions. These social constructions can then affect decisions and views of policy makers and public. Also "critical discourse analysis is interested in resisting the dominant mainstream discourses which structure an unjust and unsustainable society (Stibbe, 2018, p. 176). In this book, as I used CDA as a theoretical framework, I followed Fairclough's sociopolitical approach, Wodak's discourse-historical approach and Van Dijk' sociocognitive principles for the purpose of describing and explaining the data. Following the principles of critical discourse analysis, these texts (statements and letters) are analyzed from sociopolitical, historical and sociocognitive perspectives. By way of explanation, sociopolitical, because language features used by politicians (the President, governors and mayors) are used to trigger social change on the assumption that language itself is political, aiming to create new realties and ideologies; historical, because statements are interrelated texts, triggered by one another; and sociocognitive, because politicians draw on sociocognitive features such as metaphors for the sake of shaping the addressee's thinking of the social world.

Criticisms of CDA. Like any other approach to discourse analysis, CDA, as theory, has raised criticisms. CDA has been criticized for some methodological flaws for being open to different descriptions, and consequently to a lot of interpretations (Breeze, 2013; O'Halloran, 2010). Furthermore, Cutting (2015) reports that many criticized CDA for failing to be objective. For example, critical discourse analysts only choose texts that confirm ideological biases. Besides, CDA relies to a large extent on the analyst's perspective, which in turn undermines the objectivity of analysis and makes the analysis seem biased and go with the interests of the researchers (Breeze, 2013). Many scholars and researchers, however, to a large extent, addressed such defects. For example, one attempt to address this drawback (subjectivity) is that the analyst should "work on a basis of a variety of different data, methods, theories, and background information" (Wodak, 2014, p. 65). This Triangulation is emphasized in DHA for the purpose of minimizing the possibility of the researcher of being too subjective.

Another important strategy that can be employed to address subjectivity and limit interpretations resulting from CDA methodological analysis is using corpus analysis so that "analysts can go beyond single texts and conveniently explore quantitative patterns of ideological meaning in a large number of texts" (O'Halloran, 2010, p. 565). Using corpus analysis, the researcher can walk through a huge number of texts over a short period of time "since it is the software which reveals salience and not the analyst" (O'Halloran, 2010, p. 565). Accordingly, he argues that corpus-based CDA subjectivity to a large extent can be minimized.

The Theoretical Foundations: Ecolinguistics

An important aspect of development with CDA is applying CDA in ecolinguistics studies. With regards to the relationship between Ecolinguistics and discourse, Fill (2018) maintains that "Ecolinguistics deals with the role of language concerning the environment" (p. 3), and "ecolinguistics deals with the impact of language and discourse in describing, but also aggravating and perhaps alleviating, environmental problems" (p. 3). Stibbe (2018) argues that ecological thinkers and critics have had an important effort in protecting nature but what they do not do is conducting a detailed linguistic analysis through which stories such as ideologies, metaphor, framing and other stories we live by, are investigated to uncover how they are (re)produced and come to restructure how we think about the social world. He maintains that critical discourse analysis and cognitive linguistics are regarded as the most effective approaches for such analysis (ecolinguistic analysis). Accordingly, "if we combine linguistic approaches with the insights of environmental and ecological thinkers, then the result can be considered a form of ecolinguistics" (Stibbe, 2018, p. 499). Ecolinguistics, then, is about critiquing or exposing texts that contribute to ecological destruction (Bang & Trampe, 2014; Stibbe, 2015). In this respect, Sedlaczek (2016) asserts that this concept is used to refer to the kind of interaction between living beings and their environments revealed by discourse which in turn can add to the oriented social focus of critical discourse studies. Stibbe (2015) argues that the aim of ecolinguistics is "to encourage people to be more critical, and reject or accept discourses based not just on personal or social considerations but ecological ones too" (p. 190).

Furthermore, researchers define (e.g., Fill, 2018; Stibbe, 2018) Ecolinguistics as a variety of different approaches. However, the old approach of ecolinguistics focused on the language system while the latest approaches have focused on discourse. According to Stibbe, unlike a language system approach to ecolinguistic analysis, "a discourse approach examines how particular groups in society select particular lexical items and grammatical structures from those available from the language system, and combine them in particular ways to tell stories about the world" (Stibbe, 2018, p. 499). Thus, ecolinguistics deals "with the larger ecological systems that all life depends on" (Stibbe, 2014, p. 117). In this respect, Stibbe (2014 & 2015) argues that while CDA is particularly designed for empathy with dominated people, ecolinguistics stands with the environment including people, animals and plants, and therefore ecolinguistics considers "the larger ecological systems that all life depends on" (Stibbe, 2014, p. 117).

However, Ecolinguistics as an approach heavily relies on CDA to uncover the constructed meanings and messages signaled by linguistic features (e.g., kinds of metaphors used) that are ideologically loaded about the environment. In this regard, Verhagen (2008) identifies one of the most important functions of ecolinguistics: "to contribute to the unmasking of myths, assumptions, and ideologies that underlie the public's and scientists' notions of Nature and related issues." (p.1). This unmasking requires conducting an in-depth analysis to language (texts) used to address nature-related issues such as climate change. In Ecolinguistics, language is seen as an instrument in identifying environmental discourses (Fill, 2018; Stibbe, 2014). Although both fields (Ecolinguistics and CDA) aim at raising peoples' awareness of serious problems, "ecolinguistics advocates a specific ecological point of view, which can add to the mainly social focus of CDS" (Sedlaczek, 2016, p. 16). Stibbe is among the first scholars who drew on both approaches (ecolinguistics and CDA) to investigate environmental problems on the basis that ecological and social problems are not different. However, CDA (outlined by many scholars like Fairclough, 2014) examines texts from a political point of view to investigate how language features (metaphors), as an ideologically loaded linguistic feature, are employed by politicians to create realities or myths. According to Stibbe, 2015, when the critical discourse analyst takes an ecolinguistic approach, he/she develops a normative framework (which can be explicit or implicit) so that he/she judges language (discourses) as positive and negative based on his/her philosophy. The critical discourse analyst, therefore, situates discourse in a larger ecological or environmental system where relationships of humans with animals and plants through language are investigated.

The Theoretical Foundations: Metaphor

Early theories. Metaphor is defined as one experience or domain that is abstract or complex is described in terms of another experience or domain that is more concrete or simple (Gibbs, 1994; Lakoff & Johnson, 1980; Semino, 2008). Stibbe (2015) also defines metaphor as "a story that describes something as if it were something else" (p. 63). Similarly, Muller (2016) defines metaphor as "the transfer of a word from one place in the language to another one and the etymology of metaphor indicates this idea of transfer too" (p. 33). Lakoff and Johnson (1980) outlined and developed conceptual metaphor theory (*source and target* cross-domains mapping). They (1980) assert that metaphor can play other than being "a device of the poetic imagination and the rhetorical flourish" (p. 3). That is, they argue that "the essence of metaphor is understanding and experiencing one kind of thing in terms of another" (p. 5). Conceptual metaphor theory made researchers investigate metaphor not only as a linguistic device but also as a cognitive device. In this respect, Nerlich et al. (2002) argue that conceptual metaphors are cognitive and social devices as well, rather than rhetorical ones, realized as the result of inferential mappings between the two conceptual domains: from the source domain to the target domain. Hence, this theory (Conceptual Metaphor Theory) is concerned with the cognitive aspects of metaphor analysis (e.g., cognitive mappings). However, some scholars (e.g., Gibbs, 2017b) criticized CMT as being "relying too much on the intuitions of individual analysists.... and its failure to acknowledge the complexities in people's ongoing metaphorical experience" (Gibbs, 2017b, p. 15).

In the analysis of metaphor, Lakoff and Turner (1989) distinguished between two levels of analysis: the linguistic and conceptual metaphor and both levels complement one another. Accordingly, "metaphorical expressions in discourse are seen as linguistic reflexes of, or prompts for, conceptual structures and processes. Metaphors are not seen as mere tropes, then, but rather, the conceptual structures and processes involved in metaphor shape our thoughts and actions" (Hart, 2018, p. 188). Metaphor is seen as "a cognitive process that involves conceptualizing one thing in terms of another. The metaphorical expressions found in language are therefore deemed to be the realizations of the kinds of concepts held at the cognitive level" (Ng, 2018, p. 216). Thus, conceptual metaphors contribute to shaping or reinforcing a certain perception (Kapranov, 2017; Sullivan, 2017).

Current understandings of metaphor. Metaphor is broadly employed by speakers/writers especially in politics and media. "Successful politicians generally use metaphors, and this explains why it is of such interest to political scientists as well as linguists" (Charteris-Black, 2014, p.161). Politicians draw on metaphorical language as one way to simplify the idea (to make the topic seems much less complex than it really is) and deliver the meaning to the addressee by drawing on everyday experience. For example, many researchers (e.g., Fill, 2018; Fløttum, 2017; Nerlich et al., 2010) argue that climate change is abstract and complex and it is not easy to be grasped and understood using only nonmetaphorical (literal) language. Charteris-Black (2014) further asserts that journey, war, light, darkness, life and death are among the most prominent semantic fields that are lexicalized in American and British political metaphors. "While scientists are said to learn via analytic thinking, non-scientists are said to learn from personal experience" (Asplund, 2011, p. 2). Metaphors are, therefore, heavily utilized by nonexperts or nonscientists drawing on people's everyday experiences to conceptualize abstract issues such as climate change. According to Semino, 2008, metaphor can mediate between discourse and ideologies. That is, discourse is formulated using metaphors which is in turn can produce ideologies. Thus, "cognitive linguistic research primarily explores systematic relations in language to infer underlying linguistic, cognitive, and neural structures that motivate the creation and use of different words and linguistic expressions" (Gibbs, 2017a, p. 451). Source-target domain is among the most primary systematic relations to realize the underlying structure. "Conceptual metaphors are defined as systematic sets of correspondences, or 'mappings', across conceptual domains, whereby a 'target' domain …is partly structured in terms of a different 'source' domain" (Semino, 2008, p. 5). Therefore, the conceptual metaphor is named in the format 'TARGET DOMAIN is SOURCE DOMAIN' (Sullivan, 2017, p. 392) where it "represents a convenient summary description of the rich set of mental mappings that characterize the complex relationship between target [ ] and source [ ] domains" (Gibbs, 2017a, p. 452). In Journey metaphors, for example, the researcher investigates the linguistic surface as a primary step to reveal what ideologies that can be inferred from such expressions about the target domain. Charteris-Black (2005) says that:

> Our experience of journeys is that they are normally purposeful and goal-oriented and that different types of experiences, difficulties etc. may be encountered. Analysis of how metaphors are used to create the myths that underlie an ideology begins with identification of their source domain (p. 39).

Since metaphors are widely used in the media, it is important to highlight its role in politics and media discourse. One of the most important roles that metaphors play is that they can be employed to create rather than pass knowledge. In this regard, Fairclough (1992) argues that "when we signify things through one metaphor rather than another, we are constructing our reality in one way rather than another" (p. 194). Charteris-Black (2014) also argues that "they [metaphors] frame ways of thinking about the social world that actually constructs power relations and become political realities" (p.161). Thus, every metaphor choice that the speaker/ the writer makes is "a way of seeing the world that is favored via that choice and not others" (Johnstone, 2008, p. 54).

Metaphors also arouse emotions that can be used as the basis for evaluating political actors and actions by offering persuasive representations of social groups and social issues" (Charteris-Black, 2014, p. 161). There are several kinds of metaphors used to address climate change by politicians including War metaphor, Illness metaphor, Journey metaphor, and others (Asplund, 2011; Cohen, 2011; Fløttum, 2017). In this regard, Semino (2008) asserts that "the use of metaphor is particularly necessary in politics since politics is an abstract and complex domain of experience, and metaphors can provide ways of simplifying complexities and making abstractions accessible" (p. 90). Another reason of why metaphors are predominantly employed in politics is that the main goal of any politician is persuasion, and one strategy to do so is by drawing on metaphor (Charteris-Black, 2005), and thereby the language used by politicians become less threatening.

Metaphor has a key role in determining how the reader/listener can view the social world, and act accordingly. This proves that metaphor is an effective cognitive tool that writers/speakers use in different political and social domains to shape meanings by highlighting one aspect of reality and hiding others so that the reader/listener can perceive the world in certain ways (e.g., Lakoff & Johnson, 1980). They (Lakoff & Johnson, 1980) further maintain that when metaphor directs us to focus on one aspect of concept, "a metaphorical concept can keep us from focusing on other aspects that are inconsistent with that metaphor" (Lakoff & Johnson, 1980, p.10). Thus, "the cognitive characteristic is that a metaphor is caused by, and may cause, a shift in the conceptual system" (Charteris-Black, 2005, p. 15).

Metaphor analysis is, therefore, conducted from the point of view that metaphors bear many ideological loadings, corresponding to different perspectives that speakers/writers hold and through which dominance is (re)produced, and thereby "different metaphors have different ideological attachments" (Fairclough, 2014, p. 137). For example, "such metaphors (War metaphors) alert us that the conditions and the participants in the metaphorical battle of climate change have been radically altered and "us against climate change" has become "us against them" (Fløttum, 2017, p. 78). This to a considerable extent reflects the nature of climate change politics in the U.S.

Directionality of mappings in the conceptual metaphor. When metaphor is investigated from a cognitive perspective, there are two important domains, with one domain mapping onto the other. "CMT involves only two domains and unidirectional mappings from SOURCE to TARGET" (Sullivan, 2017, p. 394). That is, when one source domain maps onto a target domain, the target domain, in turn, does not map onto the source domain (e.g., Holyoak & Stamenkovic, 2018; Sullivan, 2017; Wolff & Gentnerb, 2011).

This kind of mapping that is involved in conceptual metaphor analysis is a one-way mapping. "The directional projection view emphasizes that in metaphor, information is projected from a familiar, often concrete base domain to a less familiar or less clear target" (Wolff & Gentnerb, 2011, p. 1456). Let's put this into practice with regards to the metaphors analyzed in the book. In journey metaphors, for example, the notions of the journey domain map onto climate change but those of the climate change domain does not. That is, the speaker draws on the journey domain, which constitutes an important aspect of everyday people's experiences, as being easy to understand. With this mapping, climate change is conceptualized as journey. But mapping does not occur from the target (the abstract) to the source domain (the concrete). This means that while the notions of journey (movement, forwardness, progress, the right path) and destination transferred from the source to target domain, no concepts are transferred from climate change onto journey because journey is called upon to facilitate and as a means to conceptualize the target domain (the politicians used this domain to trigger entailments in terms of this physical experience). This means that climate change is conceptualized in terms of journey, journey is not conceptualized in terms of climate change. As a result, the relationship is not mutual.

In turn, if a speaker/writer wants to conceptualize or describe journey using metaphors, he/she is highly unlikely to draw on climate change as a good choice to conceptualize journey. "Metaphors allow the well-developed cognitive structures from the easily understood concrete domains to be applied to the more difficult abstract domains" (Sullivan, 2017, p. 395).

The power of metaphor. Metaphors are used by speakers/writers who draw on a variety of source domains originating in our daily experiences to influence how the addressee thinks of the social world (Charteris-Black, 2014; Gibbs, 2017b; Semino, 2008; Thibodeau & Boroditsky, 2011). Due to their tremendous impact, many researchers (e.g., Van Dijk, 2003) throw light on metaphors and investigate the role that they can play. In this respect, Van Dijk (2003) maintains that metaphors can negatively influence people. That is, because of the use of such language features on the part of the writer/the speaker, the audience might adopt one opinion over the other (the opinion that the speaker/writer calls for, which serves his/her interest). One way to explain this is that "metaphor can provide a conceptual structure for a systemized ideology that is expressed in many texts and much talk" (Chilton & Schäffner, 2002, p. 29). Also, metaphors can be used not only to provide conceptual structures, but "they always have pragmatic added values, for example, to express an evaluation of the topic, to make an emotional and persuasive appeal…" (Musloff, 2016, p. 4). Thus, "metaphoric expressions, whether positive or negative with respect to their ecological origin and force (think survival vs danger) are ways of thinking and speaking that not only reflect, but also influence, our common social praxis" (Mey, 2018, p. 213). Similarly, at the linguistic level, Mey (2018) used *foe* and *friend* terms to classify metaphors used to express ideologies of politicians. She further maintains that this distinction is based on for what it is used (purpose): For example, whether it is used for explaining and facilitating a concept; or used for invading reality. "Metaphors are essential tools, needed for our survival in an environment that is not always user friendly" (Mey, 2018, p. 213). Consequently, researchers need to distinguish between metaphors in this regard (which metaphors promote protecting or destructing the environment). Positive metaphors correspond to *friends* metaphors, and, in turn, negative metaphors correspond to *foe* metaphors, according to many researchers including (Mey, 2018). However, researchers argue that "there is also a danger, for illness, war, and journey metaphors are deeply entrenched in political communication, and their transfer to climate change discourse risks turning the issue into politics as usual" (Atanasova & Koteyko, 2017, p. 84). The issue becomes more political than a problem which needs to tackled.

Components of metaphor. Based on Conceptual Metaphor Theory, which was outlined and developed by Lakoff and Johnson (1980), there are two primary domains where characteristics map from one domain onto the other one: source domain and target domain. According to Lakoff and Johnson (1980), the source domain represents the less complex and abstract than the target domain. Lakoff and Johnson (1980) and Charteris-Black (2014, 2018) maintain that the source domain is inferred (from a set of lexical items that are semantically related) rather than extracted from the data. Conceptualization is, therefore, realized through a combination and connection of lexical items belonging to a domain like journey or war. Source domain is used to make the target notion available and easy to process and grasp for the reader. Thus, by using specific items by the speaker/writer, particular characteristics are transferred, and thereby a specific perception is formed in the addressee's mind (Lakoff & Johnson, 1980). Kövecses (2016) says that the speaker/ writer has two ways when expressing target domain meanings such as progress, control, stability of structure and the like: either directly; the speaker employs a linguistic feature or an element that comes directly from the target domain, and thereby he/she speaks directly; and the other way which is indirectly; the speaker uses or borrows elements from a source domain to express the target domain, and these elements from the source to target are systematically linked by a set of mappings. This mapping is necessary to process the meaning intended by the speaker. "Metaphor comprehension entails mapping semantic and evaluative implications of the vehicle…onto the metaphor topic" (Gentner, 1983; Gibbs, 1992; Ortony, Reynolds, & Arter, 1978, cited in Ottati et al., 1999, p. 688).

Purposes of metaphor. Speakers/ writers use metaphors for many purposes according to many researchers (e.g., Charteris-Black, 2014, 2018; Semino, 2008). This book centers on metaphor as the primary linguistic and cognitive device used by the participants in their statements and letters. There are three purposes for why politicians draw on metaphor:

Framing. Entman (1993) maintains that framing is "to select some aspects of a perceived reality and make them more salient in a communicating text, in such a way as to promote a particular problem definition, causal interpretation, moral evaluation, and/or treatment recommendation for the item described" (p. 53). This means that framing is introducing the problem in specific ways that the speaker/writer desires. Thibodeau and Boroditsky (2015) maintain that "the frames will make people more likely to select policies that are congruent with the entailments of the metaphors" (p. 7). Framing is performed via using framing devices such as metaphor (Fløttum, 2017; Littlemore, 2017; Stibbe, 2017). Stibbe (2015) maintains that "metaphors are a type of framing which can be particularly powerful and vivid since they use a specific, concrete and clearly distinct frame to think about an area of life" (p.186). Sweetser (2017) also argues that metaphoric mappings (from source to target domains) shapes reasoning, with the result that we may reason about something differently based on what metaphors are used for the metaphorical mapping. The same audience might perceive two oppositional realities from two groups based on what metaphorical expressions each used.

Explanation. Climate change itself (the target domain) is abstract and complex (Fløttum, 2017). Scientific language introduced by scientists and experts on climate change is not easy to understand according to a survey conducted by pew research center in 2018. Therefore, "An increased attention to the role of language and communication is urged upon scientists by many commentators… scientific communicators are urged to adapt their language to suit the tastes, meanings and concerns of ordinary people" (Nerlich et al., 2010, p. 102). In this respect, metaphors can help facilitate comprehending complex phenomena by drawing on more concrete domains (Charteris-Black, 2014; Kövecses, 2016; Landau & Keefer, 2014; Larson, 2011; Littlemore, 2017; Raymond et al., 2013) so that people can trigger change or social action. According to many researchers, using metaphors plays an important role in explaining what is considered abstract and complex through making it look concrete and simple, and hence more available for the addressee to process. Grady (2017) maintains that "metaphors appear to be effective… when they involve reification" (p. 452). He further explains reification as "treating a phenomenon as though it were (relatively) concrete" (p. 452). Therefore, describing metaphors using metaphors from Journey, War, Cleanliness and Construction domains facilitate understanding climate change by making it less abstract. Researchers from experimental research also found out that meaning or reality constructed using figurative language is more memorable (Burgers et al., 2016).

Persuasion. Metaphors serve as means of persuasion as they direct the addressee to specific ways to think of the problem and what solutions or measures which should be implemented. This explains why metaphors are predominantly used. They function as rhetorical strategies that politicians employ to persuade their recipients that what they say is true. This persuasion is performed via appealing to the addressee's emotions and feelings (Charteris-Black, 2018). Such metaphors aim at affecting people's emotions because it is the starting stage of persuading the addressee. "Metaphor is effective in public communication because it draws on the unconscious emotional associations of words and assumed values that are rooted in cultural and historical knowledge" (Charteris-Black, 2018, p. 202). The politician aims to persuade the addressee that, for example, climate change is threatening, and therefore efforts need to be intensified.

The politician's main concern is to construct this meaning as reality or knowledge.

Patterns of metaphor. Charteris-Black (2016) and Semino (2008) argue that metaphors are used in a complex or diverse manner in political discourse. Many scholars (e.g., Charteris-Black, 2016; Dorst, 2017; Kövecses, 2016; Semino, 2008) identified different types of textual patterns of metaphors. Among the most common ones are:

Repetition. Repetition exists in this data where "the text of stretch of discourses contains several instances of the same metaphorical expression" (Dorst, 2017, p. 179). Several metaphorical expressions frequently occur in a text or a number of texts about a topic. An important function for repetition of metaphors by the speaker/writer is that it facilitates comprehension since the addressee processes repeated metaphors in much less time than new ones (Phillips, 2017). Because texts do not exist in a vacuum, the possibility of repeating the same metaphorical expression throughout a text or a number of texts by the speaker/writer is high (Fairclough, 2014). According to Semino (2008), repetition indicates the metaphorical productivity of particular concepts. She maintains that repetition is an important patterning because it contributes to shaping realities in texts and it reveals correspondence of source domains to target domains, and therefore inferences are easier to be made on the part of the addressee.

Extension. Extension "is where a series of semantically related metaphor vehicles describe the same topic" (Charteris-Black, 2016, p. 162). It is also defined by Semino (2008) as occurring when "several metaphorical expressions belonging to the same semantic field or evoking the same source domain are used in close proximity to one another in relation to the same topic, or to elements of the target domain" (p. 25). She explains this (why several metaphorical expressions of the same semantic field occur in proximity when addressing a problem) from a cognitive perspective. In this respect, she maintains that "once a source domain is active, it may be exploited more than once as a source of metaphorical expressions within relatively short stretches of text" (Semino, 2016, p. 220). All of such expressions, therefore, can "contribute to a single coherent image" (Charteris-Black, 2017, p.162) since "the source domains in conceptual metaphors are primarily image-schematic" (Gibbs, 2017a, p. 452). This textual pattern differs from mixing in that in extension, metaphors should be drawn from the same source domain. Simply put, in this pattern (extension), there are different metaphorical expressions but belong to the same source domain for the purpose of describing one target domain/ topic (Dorst, 2017).

***Mixing*.** Mixing is when two metaphorical expressions from two or more different source domains describe the same topic (the same target domain) existing in a single clause "cities will continue to lead the way in moving forward in protecting our residents" (from the mayors' letter to the president). Kimmel (2010) states that metaphor clusters fulfil three kinds of functions: "First, they are attention-grabbing..... Second, clusters seem to occur where the action is. Finally, metaphor clusters connect and dynamize discourse" (p. 98). Many scholars discussed the notion of mixing (Kimmel, 2010; Lakoff & Johnson, 1980; Semino, 2008). The linguistic environment is important in determining this type of pattern, which is represented and identified by the clause borders. They set out grammatical boundaries for mixed metaphors: occur across adjacent clauses within the same sentence. "The mixing of metaphors creates a semantic inconsistency on the level of literal meaning" (Muller, 2016, p. 41), and one of the scenarios of this clash or inconsistency is that it can be subsumed under the same or different source domains (Semino, 2016). This is what John Searle pointed out in his book by saying that literal meaning first hypothesis. Identifying mixed metaphors has been beneficial in understanding the topic conceptualized by such mixed metaphors. In medical discourse, for example, Charteris-Black (2016) says that when the patient feels severe pain, he/she uses metaphors from different domains. That is, through experimental research he found that a patient draws on source domains that are widely different to describe how severe and complex the problem is. "The greater semantic divergence of metaphor source domains, the more intense the embodied experience of pain" (Charteris-Black, 2016, p.155). Thus, an individual target domain can be conceptualized or conceived through different and unrelated source domains (Kövecses, 2010, 2016) for the purpose of helping the speaker/writer to better and more comprehensively describe or conceptualize the target domain to the addressee.

Familiarity with source domain. Familiarity represents the extent of knowledge that the addressee has about a (source) domain. Landau and Keefer (2014) maintain that "many source concepts derive from familiar sensorimotor experiences such as losing one's balance, firmly grasping objects, moving toward destinations, and avoiding physical filth. Others represent stereo typed cultural knowledge (e.g., how buildings are constructed..)" (p. 464). Boroditsky (2017) and Thibodeau and Boroditsky (2011) explain people's familiarity through this metaphor: Crime is virus. They argue that there is a stronger consensus among people on how to deal with virus than with crime, which should be through social treatment rather than other means like enforcement.

In this book, the source domains which the politicians drew on to conceptualize the target domain are common. People are more familiar with these domains than the target domain of climate change. That is, they represent physical experiences drawn from our real and everyday lives. By way of explanation, journey constitutes an important aspect of everyday experience in people's lives since people move from one place to another to reach different destinations on a daily basis. This also explains why politicians heavily draw on journey metaphors. For war, people might not go through battles, but war constitutes an important aspect of history of many nations around the world, including the U.S. Also, people's familiarity with this domain is through watching and hearing about wars, battles and crises on TV and radio (Flusberg et al., 2018). This explains why metaphors of War are very widely used in different social, religious and political domains, like war on poverty, war on cancer, war on drugs and war on crime (Flusberg et al., 2017). For the construction domain, politicians draw on this domain as building is always present in people' lives and most people are fully aware of how building is performed and the importance of building. Construction implies a set of processes that workers with different specialties need to go through to finally have a well-structured building. Cleanliness is here called upon by the politicians since the addressee realizes the importance of cleanliness in his/her everyday life, "cleanliness can be considered a basic domain of experience, due to the human need for being clean in order to preserve health and basic hygiene" (Pavlovic, 2012, p. 29). Neglecting cleanliness thus causes a lot of diseases that might be life-threatening. This can be one reason why politicians draw on this semantic domain.

These metaphors, therefore, from such familiar domains used by politicians, can facilitate the addressee's perception of the politicians' meanings about such abstract topics as climate change. "Metaphors and analogies are known to help people apply their understanding of familiar and/or concrete topics to new/or more abstract ones" (Grady, 2017, p. 447). Familiarity is necessary since it enables the addressee to make comparisons (mappings discussed in the analysis chapter) between the source and target domain. Colston and Gibbs (2017) said that many experimental studies on metaphor processing (like reading-time or word-fragment completion), such as Giora 2012, found that processing familiar metaphors takes much less time than unfamiliar ones. That is, for familiar ones, the semantic and metaphorical meanings are simultaneously activated. For unfamiliar metaphors, however, semantic meaning is processed (where the addressee perceives a clash on the semantic level) before metaphorical one because "psycholinguistic research has shown that when the expected word fails to appear and an unexpectedly different one **takes** its place, comprehension is delayed ever so slightly" (Phillips, 2017, p. 225). These findings (Giora, 2012; Phillips, 2017) of the experimental study prove that familiarity with metaphors facilitates perceiving reality as well as the complexity of the target domain. Moreover, an earlier experimental study by Bowdle and Gentner (2005) agrees with Giora's (2012) findings that more familiar (conventional) metaphors are processed much easier and take much less time to be comprehended in comparison with metaphors that can be less familiar (novel) metaphors.

Main Types of Research Investigating Metaphor

As this book analyzes, and focuses on, metaphor, it is worth shedding light on types of research investigating this feature. Based on reviewing studies on environmental issues examining metaphors, it can be argued that research into metaphor falls into three main areas. They are distinctive by the methods each uses. However, although these types of research each employ different research tools, they try to address the same basic research questions (like what metaphors do speakers/writers use to influence people's opinions? how is an environmental problem conceptualized? and how does metaphor shape our attitudes towards the problem?).

CDA oriented analysis. In this type of research, analysis of language features is conducted from CDA and cognitive perspectives to explain how the addressee reason about the social world (through the speaker/writer's use of linguistic features). Critical metaphor analysis is among the most recent type of this research where the researcher draws on CDA models to explain how the speaker/writer constructs the social world. Charteris-Black (2011, 2014 & 2018) developed critical metaphor analysis, which is CDA-based, to explain how speakers/writers use metaphors to trigger change from the point of the view that metaphors influence the way we think of the world. His approach focuses on conceptual metaphors as the basis of analysis at all the stages. This approach enriches analysis with cognitive aspects by highlighting metaphors as cognitive features responsible for forming attitudes.

Corpus analysis. There are many studies (e.g., Raffaelli & Katunar, 2016; Sardinha, 2011; Wills, 2017) in which researchers used corpus analysis to investigate metaphor. Among the programs used in corpus-linguistics studies are the Metaphor Candidate Identifier and Wmatrix software. Most of these studies integrate corpus analysis with CDA since CDA is qualitative research and is open to a lot of explanations, CDA researchers draw on corpus analysis as one way to limit these explanations by referring to, and judging based on, statistics rather than assumptions. Both kinds of research are complementary to one another "corpus analysis is best combined with CDA" (Wills, 2017, p. 220). That is, she (Wills, 2017) argues that while CDA needs quantitative findings to minimize subjectivity of the findings, corpus analysis also requires explaining the quantitative findings using qualitative research.

Experimental analysis. Research especially in social psychology and psycholinguistics employs experimental methods and focus groups to investigate how metaphors influence the addressee's perceptions of the social world and how the addressee processes metaphors. For example, Flusberg et al. (2017) investigated if war or race metaphors have more powerful effects in developing a sense of urgency in the addressee. In this type of analysis, researchers (e.g., Boroditsky, 2017; Thibodeau et al., 2017; Thibodeau & Boroditsky, 2011; Thibodeau & Boroditsky, 2015) employ measuring tools, such as tests, surveys, questionnaire and among others to measure participants' perceptions and if one type of metaphor can develop a stronger urgency in the participant than the other.

References

Amossy, R. (2018). Understanding Political Issues Through Argumentation Analysis. In R. Wodak & B. Forchtner (Eds.), *The Routledge Handbook of Language and Politics* (pp. 262–275). Routledge.

Anastassov, V. (2018). *Power and Truth in Political Discourse: Language and Ideological Narratives*. Cambridge Scholars Publishing.

Asplund, T. (2011). Metaphors in Climate Discourse: An Analysis of Swedish Farm Magazines. *JCOM, 10*(04), 1–8.

Atanasova, D., & Koteyko, N. (2017). Metaphors in Online Editorials and Op-ed About Climate Change, 2006-2013: A Study of Germany, the United Kingdom, and the United States. In K. Fløttum (Ed.), *The Role of Language in the Climate Change* (pp. 71–89). Routledge.

Bang, J. C., & Trampe, W. (2014). Aspects of an Ecological Theory of Language. *Language Sciences, 41*, 83–92.

Boroditsky, L. (2017). The Function of Metaphor Framing, Deliberate or Otherwise, in a Social World. *Metaphor and the Social World, 7*(2), 270–290.

Bowdle, F. B., & Gentner, D. (2005). The Career of Metaphor. *Psychological Review, 112*(1), 193–216.

Breeze, R. (2013). Critical Discourse Analysis and Its Critics. *Pragmatics, 21*(4), 493–525.

Burgers, C., Konijn, E. A., & Steen, G. J. (2016). Figurative Framing: Shaping Public Discourse Through Metaphor, Hyperbole, and Irony. *Communication Theory, 26*(4), 410–430.

Carvalho, A. (2018). Climate Change and the Socio-ecological Crisis. In A. F. Fill & H. Penz (Eds.), *The Routledge Handbook of Ecolinguistics* (pp. 487–499). Routledge.

Charteris-Black, J. (2005). *Politicians and Rhetoric. The Persuasive Power of Metaphor*. Palgrave Macmillan.

Charteris-Black, J. (2011). *Politicians and Rhetoric. The Persuasive Power of Metaphor*. Palgrave Macmillan.

Charteris-Black, J. (2014). *Analyzing Political Speeches: Rhetoric, Discourse and Metaphor* (1st ed.). Palgrave Macmillan.

Charteris-Black, J. (2016). The Dull Roar and the Burning Barbed Wire Pantyhose: Complex Metaphor in Accounts of Chronic Pain. In R. W. Gibbs (Ed.), *Mixing Metaphor* (pp. 155–176). John Benjamin publishing Company.

Charteris-Black, J. (2018). *Analyzing Political Speeches: Rhetoric, Discourse and Metaphor* (2nd ed.). Palgrave Macmillan.

Chilton, A. P., & Schäffner, C. (2002). *Politics as Text and Talk: Analytic Approaches to Political Discourse*. John Benjamins Publishing Company.

Cohen, M. J. (2011). Is the UK Preparing for War? Military Metaphors, Personal Carbon Allowances and Consumption Rationing in Historical Perspectives. *Climate Change, 104*, 199–222.

Colston, H., & Gibbs, W. (2017). Metaphor Processing. In E. Semino & S. Demjen (Eds.), *The Routledge Handbook of Metaphor and Language* (pp. 457–471). Routledge.

Cutting, J. (2015). *Pragmatics: A Resource Book for Students* (3rd ed.). Routledge.

Dorst, A. (2017). Textual Patterning of Metaphor. In E. Semino & Z. Demjen (Eds.), *The Routledge Handbook of Metaphor and Language* (pp. 178–192). Routledge.

Entman, R. M. (1993). Framing: Toward Clarification of a Fractured Paradigm. *Journal of Communication, 43*(4), 51–58.

Fairclough, N. (1989). *Language and Power* (1st ed.). Pearson Education Limited.

Fairclough, N. (1992). Discourse and Text: Linguistic and Intertextual Analysis within Discourse Analysis. *Discourse & Society, 3*(2), 193–217.

Fairclough, N. (2001). *Language and Power* (2nd ed.). Pearson Education Limited.

Fairclough, N. (2014). *Language and Power* (3rd ed.). Routledge.

Fairclough, N., & Wodak, R. (1997). Critical Discourse Analysis. In T. VanDijk (Ed.), *Discourse as Social Interaction* (pp. 258–284). Sage.

Fill, F. A. (2018). Introduction. In A. F. Fill & H. Penz (Eds.), *The Routledge Handbook of Ecolinguistics* (pp. 1–7). Routledge.

Fill, F. A., & Muhlhausler, P. (2001). *The Ecolinguistics Reader: Language, Ecology and Environment*. Continuum.

Fill, F. A., & Penz, H. (2018). Ecolinguistics in the 21st Century: New Orientations and Future Directions. In A. F. Fill & H. Penz (Eds.), *The Routledge Handbook of Ecolinguistics* (pp. 437–443). Routledge.

Fløttum, K. (2017). Willingness of Action. In K. Fløttum (Ed.), *The Role of Language in the Climate Change Debate* (pp. 113–129). Routledge.

Flusberg, S., Matlock, T., & Thibodeau, P. (2017). Metaphors for the War (or race) Against Climate Change. *Environmental Communication, 6*, 769–783.

Flusberg, S., Matlock, T., & Thibodeau, P. (2018). War Metaphors in Public Discourse. *Metaphor and Symbol, 33*(1), 1–18.

Gee, J. P. (1999). *An Introduction to Discourse Analysis: Theory and Method.* Routledge.

Gee, J. P. (2014). *An Introduction to Discourse Analysis: Theory and Method.* Routledge.

Gibbs, W. R. (1994). *The Poetics of Mind: Figurative Thought, Language, and Understanding.* Cambridge University Press.

Gibbs, W. R. (2017a). Embodiment. In B. Dancygier (Ed.), *The Cambridge Handbook of Cognitive Linguistics* (pp. 449–462). Cambridge: Cambridge University Press.

Gibbs, W. R. (2017b). *Metaphor Wars: Conceptual Metaphors in Human Life.* Cambridge: Cambridge University Press.

Grady, J. (2017). Using Metaphor to Influence Public Perceptions and Policy: How Metaphors Can Save the World. In E. Semino & S. Demjen (Eds.), *The Routledge Handbook of Metaphor and Language* (pp. 443–454). Routledge.

Halliday, M. A. K. (2001). New Ways of Meaning: The Challenge to Applied Linguistics. In F. A. Fill & P. Muhlhausler (Eds.), *The Ecolinguistics Reader: Language, Ecology and Environment* (pp. 175–202). Continuum.

Hart, C. (2018a). Event-frames Affect Blame Assignment and Perception of Aggression in Discourse on Political Protests: An Experimental Case Study in Critical Discourse Analysis. *Applied Linguistics, 3*, 400–421.

Hart, C. (2018b). Cognitive Linguistic Critical Discourse Studies: Connecting Language and Image. In R. Wodak & B. Forchtner (Eds.), *The Routledge Handbook of Language and Politics* (pp. 187–201). Routledge.

Hart, C., & Cap, P. (2014). Introduction. In C. Hart & P. Cap (Eds.), *Contemporary Critical Discourse Studies* (pp. 1–15). Bloomsbury.

Holyoak, J. K., & Stamenkovic, D. (2018). Metaphor Comprehension: A Critical Review of Theories and Evidence. *Psychological Bulletin, 6*, 641–671.

Johnstone, B. (2008). *Discourse Analysis* (2nd ed.). Blackwell Publishing.

Kapranov, O. (2017). Conceptual Metaphors Associated with Climate Change in Corporate Reports in the Fossil Fuels Market: Two Perspectives from the United States and Australia. In K. Fløttum (Ed.), *The Role of Language in the Climate Change Debate* (pp. 90–109). Routledge.

Kimmel, M. (2010). Why We Mix Metaphors (and mix them well): Discourse Coherence, Conceptual Metaphor, and Beyond. *Journal of Pragmatics, 42*, 97–115.

Kövecses, Z. (2010). *Metaphor: A Practical Introduction* (2nd ed.). Oxford University Press.

Kövecses, Z. (2016). A View of Mixed Metaphor within a Conceptual Metaphor Theory Framework. In R. W. Gibbs (Ed.), *Mixing Metaphor* (pp. 3–15). John Benjamins Publishing Company.

Lakoff, G., & Johnson, M. (1980). *Metaphors We Live By*. The University of Chicago Press.

Lakoff, G., & Turner, M. (1989). More than cool reason: A field guide to poetic metaphor, Chicago: University of Chicago Press.

Landau, M. J., & Keefer, L. A. (2014). This Is Like That: Metaphors in Public Discourse Shape Attitudes. *Social and Personality Psychology Compass, 8*(8), 463–473.

Larson, B. (2011). *Metaphors for Environmental Sustainability: Redefining Our Relationship with Nature*. Yale University Press.

Leeuwen, T. (2018). Legitimation and Multimodality. In R. Wodak & B. Forchtner (Eds.), *The Routledge Handbook of Language and Politics* (pp. 218–232). Routledge.

Littlemore, J. (2017). Metaphor Use in Educational Contexts: Functions and Variations. In E. Semino & Z. Demjen (Eds.), *The Routledge Handbook of Metaphor and Language* (pp. 283–295). Routledge.

Mey, J. (2018). The Pragmatics of Metaphor: An Ecological View. In A. F. Fill & H. Penz (Eds.), *The Routledge Handbook of Ecolinguistics* (pp. 211–223). Routledge.

Muhlhausler, P. (2018). What Creolistics Can Learn from Ecolinguistics. In A. F. Fill & H. Penz (Eds.), *The Routledge Handbook of Ecolinguistics* (pp. 135–148). Routledge.

Mulderrig, J. (2011). The Grammar of Governance. *Critical Discourse Studies, 8*(1), 45–68.

Muller, C. (2016). Why Mixed Metaphors Make Sense. In R. W. Gibbs (Ed.), *Mixing Metaphor* (pp. 31–56). John Benjamins publishing Company.

Musloff, A. (2016). *Political Metaphor Analysis: Discourse and Scenarios*. Bloomsbury Publishing.

Nerlich, B., Hamilton, A. C., & Rowe, V. (2002). Conceptualizing Foot and Mouth Disease: The Socio-cultural Role of Metaphors, Frames and Narratives. *Metaphorik.de, 1*, 90–108.

Nerlich, B., Koteyko, N., & Brown, B. (2010). Theory and Language of Climate Change Communication. *Climate Change, 1*, 97–110.

Ng, C. (2018). Metaphor. In J. Flowerdew & J. Richardson (Eds.), *The Routledge Handbook of Critical Discourse Studies* (pp. 215–227). Routledge.

O'Halloran, K. (2010). How to Use Corpus Linguistics in the Study of Media Discourse. In A. O'Keeffe & M. McCarthy (Eds.), *The Routledge Handbook of Corpus Linguistics* (pp. 563–577). Routledge.

Ottati, V., Rhoads, S., & Graesser, C. A. (1999). The Effect of Metaphor on Processing Style in a Persuasion Task: A Motivational Resonance Model. *Journal of Personality and Social Psychology, 77*(4), 688–697.

Pavlovic, T. (2012). Cleanliness Is Next to Godliness? Clean and Dirty Metaphors in English and Serbian. *ELALT, 1*, 27–40.

Phillips, P. (2017). Conventional and Novel Metaphors in Language. In E. Semino & Z. Demjen (Eds.), *The Routledge Handbook of Metaphor and Language* (pp. 219–232). Routledge.

Priest, S. (2016). *Communicating Climate Change: The Path Forward.* Palgrave Studies in Media and Environmental Communication.

Raffaelli, I., & Katunar, D. (2016). A Discourse Approach to Conceptual Metaphors: A Corpus-based Analysis of Sports Discourse in Croatian. *Studia Linguistica Universitatis Iagellonicae Cracoviensis, 133*, 125–147.

Raymond, M. C., Singh, G. G., Benessaiah, K., Bernhardt, J. R., Levine, J., Nelson, H., Turner, N. J., Norton, B., Tam, J., & Chan, M. A. K. (2013). Ecosystem Services and Beyond: Using Multiple Metaphors to Understand Human-environment Relationships. *BioScience, 63*(7), 536–546.

Reisigl, M. (2017). The Discourse-Historical Approach. In J. Flowerdew & J. E. Richardson (Eds.), *The Routledge Handbook of Critical Discourse Studies* (pp. 44–59). Routledge.

Reisigl, M., & Wodak, R. (2008). The Discourse-historical Approach (DHA). *Methods of Critical Discourse Analysis, 1*, 87–119.

Rogers, R. (2011). *An Introduction to Critical Discourse Analysis in Education* (2nd ed.). Routledge.

Sardinha, B. T. (2011). Metaphor and Corpus Linguistics. *RBLA, Belo Horizonte, 11*(2), 329–360.

Sedlaczek, A. (2016). Representations of Climate Change in Documentary Television: Integrating an Ecolinguistic and Ecosemiotic Perspective into a Multimodal Critical Discourse Analysis. *Language & Ecology, 1*, 1–19.

Semino, E. (2008). *Metaphor in Discourse.* Cambridge University Press.

Semino, E. (2016). A Corpus-based Study of Mixed Metaphor as a Metalinguistic Comment. In R. W. Gibbs Jr. (Ed.), *Mixing Metaphor* (pp. 203–220). John Benjamins Publishing Company.

Stibbe, A. (2014). An Ecolinguistic Approach to Critical Discourse Studies. *Critical Discourse Analysis, 11*, 117–128.

Stibbe, A. (2015). *Ecolinguistics: Language, Ecology, and the Stories We Live By*. Routledge.

Stibbe, A. (2017). Critical Discourse Analysis and Ecology. In J. Flowerdew & J. E. Richardson (Eds.), *The Routledge Handbook of Critical Discourse Studies* (pp. 497–509). Routledge.

Stibbe, A. (2018). Positive Discourse Analysis: Rethinking Human Ecological Relationship. In A. F. Fill & H. Penz (Eds.), *The Routledge Handbook of Ecolinguistics* (pp. 165–178). Routledge.

Sullivan, K. (2017). Conceptual Metaphor. In B. Dancygier (Ed.), *The Cambridge Handbook of Cognitive Linguistics* (pp. 385–406). Cambridge University Press.

Sum, N., & Jessop, B. (2013). *Towards a Cultural Political Economy: Putting Culture in Its Place in Political Economy*. Edward Elgar Publishing.

Sweetser, E. (2017). Conceptual Mappings. In B. Dancygier (Ed.), *The Cambridge Handbook of Cognitive Linguistics* (pp. 379–384). Cambridge University Press.

Thibodeau, P. H., & Boroditsky, L. (2011). Metaphors We Think with: The Role of Metaphor in Reasoning. *PLoS ONE, 6*(2), e16782. https://doi.org/10.1371/journal.pone.0016782

Thibodeau, P. H., & Boroditsky, L. (2015). Measuring Effects of Metaphor in a Dynamic Opinion Landscape. *PLoS ONE, 10*(7), 1–22.

Thibodeau, P. H., Hendricks, R. K., & Boroditsky, L. (2017). How Linguistic Metaphor Scaffolds Reasoning. *Trends in Cognitive Sciences, 21*(11), 852–863.

Van Dijk, T. (1993). Principles of Critical Discourse Analysis. *Discourse& Society, 4*(2), 249–283.

Van Dijk, T. (1995). Aims of Critical Discourse Analysis. *Japanese Discourse, 1*, 17–27.

Van Dijk, T. (1997). *Discourse as Social Interaction*. Sage.

Van Dijk, T. (2003). The Discourse-knowledge Interface. In G. Weiss & R. Wodak (Eds.), *Critical Discourse Analysis: Theory and Interdisciplinarity* (pp. 85–109). Palgrave Macmillan.

Van Dijk, T. (2008). *Discourse and Context: A Sociocognitive Approach*. Cambridge University Press.

Van Dijk, T. (2009). *Society and Discourse: How Social Contexts Influence Text and Talk*. Cambridge University Press.

Van Dijk, T. (2014). Discourse and Knowledge. In P. Gee & M. Handford (Eds.), *The Routledge Handbook of Discourse Analysis* (pp. 587–603). Routledge.

Van Dijk, T. (2015). Critical Discourse Analysis. In D. Tannen, H. E. Hamilton, & D. Shiffrin (Eds.), *The Handbook of Discourse Analysis* (pp. 466–485). Wiley Blackwell.

Van Dijk, T. A. (2018). Sociocognitive Discourse Studies. In J. Richardson & J. Flowerdew (Eds.), *The Routledge Handbook of Critical Discourse Studies* (pp. 26–43). Routledge.

Verhagen, C. F. (2008). Worldviews and Metaphors in the Human-nature Relationship: An Ecolinguistic Exploration Through the Ages. *Language & Ecology, 2*(3), 1–15.

Wills, R. (2017). Taming the Climate? Corpus Analysis of Politicians' Speech on Climate Change. *Environmental Politics, 26*(2), 212–231. https://doi.org/10.1080/09644016.2016.1274504

Wodak, R. (2001). What CDA is about- A summary of its history, important concepts and its developments. In W. R., & M. Meyer (Eds.), *Methods of Critical Discourse Analysis* (pp. 1–13). London: Sage Publications.

Wodak, R. (2008). Discourse-historical Approach. In R. Wodak & M. Meyer (Eds.), *Methods of Critical Discourse Analysis* (pp. 87–121). Sage.

Wodak, R. (2014). Politics as Usual: Investigating Political Discourse in Action. In J. P. Gee & M. Handford (Eds.), *The Routledge Handbook of Discourse Analysis* (pp. 525–540). Routledge.

Wodak, R., & Meyer, M. (2008). Critical Discourse Analysis: History, Agenda, Theory and Methodology. In R. Wodak & M. Meyer (Eds.), *Methods of Critical Discourse Analysis* (pp. 1–33). Sage.

Wodak, R., & Meyer, M. (2016). *Methods of Critical Discourse Studies* (3rd ed.). Sage.

Wolff, P., & Gentnerb, D. (2011). Structure-Mapping in Metaphor Comprehension. *Cognitive Science*, 35, 1456–1488.

4

Methods and Strategies of Sampling

The fourth chapter presents methods and strategies for sampling. It presents the politicians (the president, senators, governors and mayors) whose data was analyzed in this book. It includes all information needed about these politicians. This chapter also explains the methods used for analyzing metaphor, including identifying and interpreting metaphors and cognitive frameworks (mapping and opposition) which are used so that metaphor is explained at the linguistic and cognitive level as well.

Methods

The politicians from whom the data was collected made two opposing arguments. The president and many politicians supported the U.S. withdrawal from the Paris agreement. Other politicians, however, argued against this withdrawal.

To examine this data, I applied the analysis at two levels: linguistic and cognitive. At the linguistic level, metaphorical expressions are studied. I identified metaphorical expressions (linguistic realizations) used by the speakers, judging what counts as a metaphor, manually through source (semantic) domains. According to Charteris-Black (2014 & 2018) and

O. K. Al-Shboul, *The Politics of Climate Change Metaphors in the U.S. Discourse*,
https://doi.org/10.1007/978-3-031-19016-2_4

many other scholars, identifying what counts as a metaphor is not an easy task due to the fact that "metaphors are not inherent in words, but arise from how words are used and understood" (Charteris-Black, 2014, p. 178). Consequently, "some linguistically based approaches start with a working definition but many studies rely on intuition" (Charteris-Black, 2018, p. 206). Having identified those expressions, I extracted them from the data (statements and letters) so that I can move to the next stage which involves classifications and categorizations of these metaphors extracted. This stage focuses on how metaphors are classified, organized, and arranged. This means that source domains (categories) are realized from the metaphorical expressions used by politicians. Accordingly, after I had extracted all the metaphorical expressions out of the data, I placed them in groups based on the semantic content of these expressions (source domain). Then, each group is titled with a category (a conceptual metaphor). This category is realized based on the meaning of those expressions existing in each group. For example, all the metaphors to do with movement, progress, motion or trip are placed in one group and given a journey title. I borrowed these metaphor categories from some linguistic studies on environmental problems, like Semino, 2005; Charteris-Black, 2014, 2018; Fløttum, 2017. After that, I conducted a quantitative analysis where I counted the metaphorical expressions in each group and found the proportion of the frequency of each group out of the total metaphors existing in all groups in order to find out what the most common metaphors that are used by the politicians.

At the level of the cognitive analysis, where I indicated how the addressee can perceive realities that the politicians make, I analyzed conceptual metaphors using cognitive framework 2018, I used Lakoff's cross-domain mapping (the source and target domains of metaphors). I also developed another cognitive framework, which is *the metaphor opposition* I borrowed this term *the metaphor opposition* from Semino's work (2008) where she used this framework to make a distinction between contrasting metaphors (e.g., metaphors of life versus metaphors of death). In this book, however, it is used to make a distinction between positive (favored) and negative (unfavored) elements of a conceptual metaphor (e.g., journey) for the purpose of explaining how the addressee can be directed (through metaphors) to perceive realities and act accordingly from a cognitive perspective. In this regard, it can be argued that the concepts making up the favored part of the reality encourage the addressee to adopt what is called for and the unfavored (oppositional) concepts, in turn, serve as a warning, pushing him/her to adopt the reality and act accordingly. Consequently, positive and negative aspects both push the addressee to comply with (support and follow) the politician's argument. The opposition theory claims then that each metaphor involves concepts that can lie in two opposing side (e.g., positive and negative or favored and unfavored). These concepts arise from the semantic content of the target domain. They are regarded the most prominent concepts of that domain.

Following Charteris-Black (2014 & 2018), Fairclough (2014) and Semino (2008) who developed and employed critical discourse analysis-based approaches to analyze discourse especially that which pertains to politics, I analyzed metaphors as linguistic and cognitive features used by the participants to construct realities about climate change. In this book, I adopted and developed Charteris-Black (2014 & 2018)'s *critical metaphor methodology*, which is based on CDA theory. This framework pays special attention to metaphor due to the fact that metaphors have an important role in (re)shaping knowledge as argued by many scholars (e.g., Charteris-Black, 2005, 2014 & 2018; Lakoff & Johnson, 1980). Gibbs (2017b) maintain that "the aim of CMA is to broaden the scope of metaphor analysis to acknowledge the ideological, social, and political forces that directly influence people's choices of metaphorical words in context" (p.159). All metaphors in this book are based on physical aspects drawn from human experience, where these aspects map onto climate change.

This book adopts and develops this analytical framework (critical metaphor analysis) that can be applied to analyze language used to construct knowledge to show how CDA can be integrated into cognitive linguistics, where language is a social practice used to constitute the social world, including social relations, structures, actors and so on, as argued by Fairclough (2014) and Gee (2014).

This approach is adopted, in particular, for three mains reasons: First, this framework helps the researcher define what metaphorical expressions that speakers/writers use in their language to convince the audience of their arguments. Second, it enriches linguistic analysis with cognitive aspects of texts since it analyzes language in terms of cross-domain mapping (source and target domains), focusing on sociocognitive features that can construct realities. Third, this framework is specifically designed and developed for metaphor analysis, and thereby it enables the researcher to look into this language feature more deeply than any other framework. To do so, besides to the source-target mappings, I analyzed metaphors using metaphor opposition (I borrowed this term from Semino, 2008), where I offered a cognitive analysis of not only how the addressee is directed to perceive realities through metaphors but also how metaphors shaping these realities push him/her to take action accordingly (through the favored and unfavored elements of realities shaped by metaphors).

Charteris-Black's (2014 & 2018) framework consists of four main stages, with each stage requiring a level of analysis.

1. Contextual analysis: at this stage, I identified the context in which texts containing metaphors are produced, including information concerning the speakers, the time period, the number of speeches. At this stage, I also identified what texts are selected. There are two types of context: 1. linguistic context 2. social context. According to Charteris-Black (2014 & 2018) as explained in the methodology chapter, when analyzing a text/texts using critical metaphor approach, analysis of speech circumstances should be first introduced, as the first stage of this analytical framework. These speech circumstances include investigating situational and cognitive circumstances of the participants so that I can then analyze the participants' ideologies (the governors' and mayors') in accordance with these circumstances, especially situational and cognitive circumstances. For the situational circumstances in which the texts were produced, Van Dijk (2008, 2009) distinguishes between two types of contexts: verbal context and social situation. The first type of contexts is concerned with surrounding words and phrases within a discourse (the linguistic environment). The other type of contexts is the social situation of a discourse (e.g., when, why and how social actors produce discourse) where properties of related texts and knowledge are important parts of the context. In this book, the focus is on the two types of contexts: linguistic and social. That is, meta-

phorical expressions are investigated in terms of linguistic and social context. This book attempts at offer explanations for using metaphors. Therefore, social context is important so that the problem investigated from not only the linguistic but also social aspects, offering a more comprehensive explanation of how metaphorical expressions affect attitudes, based on the cognitive theory (conceptual metaphor theory).

These two types of contexts (verbal context and social) are discussed in the situational circumstances as the first stage: I collected metaphors not from only one person but from many politicians representing a political group (the governors, senators & mayors) at the state and local level. This sample consists of seventeen statements made by 17 governors and a letter headed to the President by 12 governors as well as a statement by 407 mayors and a letter by 63 mayors, who all opposed the President when he made a decision that he would withdraw the U.S. from the Paris accord in 2017. As a result, those texts (the statements and the letters) were produced on this occasion. However, his decision made many reactions not only from the governors and mayors but from many politicians and journalists from inside and outside the country. His announcement shocked many leaders from around the world as it was surprising that the United States which is a world leading member in tackling climate change withdraws from this important agreement. For cognitive circumstances, which is the second type of circumstances, the politicians are political leaders who discursively (socially and politically) constructed climate change in terms of what the problem is and how it can be tackled. The topic is debated in social and political contexts by the politicians (the governors and mayors). They tried to persuade the President to change his decision and remain the U.S. in the Paris climate agreement. They also tried to encourage the audience to call the federal government and policy makers for dealing with climate change. The language investigated in this book is political rather than scientific (e.g., scientific reports issued by scientists). Unlike scientific language, the political and social language is based on the speaker'/writer's personal experiences and perspectives formulated and expressed via drawing on different source domains such as Journey and War (Asplund, 2011), with the result that realities formed are of social and political nature but based on scientific proofs.

2. Metaphor identification: I identified metaphorical expressions (linguistic realizations) used by the speakers (deciding what counts as a metaphor) manually through source (semantic) domains. According to Charteris-Black (2014 & 2018) and many other scholars, identifying what counts as a metaphor is not an easy task. "in deciding what counts as a metaphor, studies vary in how far they systematically apply a definition of metaphor. Some linguistically based approaches start with a working definition but many studies rely on intuition" (Charteris-Black, 2018, p. 206). He further explained that metaphors can be classified into groups based on many criteria including source domain and target domain. Thus, to identify metaphors in the data of this book, I first highlighted all the expressions that can count as metaphors in the data (the statements and the letters). Second, I extracted all of these expressions from these statements and letters so that I can move to the next stage which involves classifications and categorizations of these metaphors extracted.
3. Metaphor interpretation: This stage focuses on how metaphors are classified, organized, and arranged. "Metaphors are not inherent in words, but arise from how words are used and understood" (Charteris-Black, 2014, p. 178). This means that source domains (categories) themselves are not mentioned in the data, but they are realized from the metaphorical expressions used by politicians. To do so, after I had extracted all the metaphorical expressions out of the data, I placed them in groups based on the semantic content of these expressions (source domain). Then, each group is titled with a category (a conceptual metaphor). This category is realized based on the meaning of those expressions existing in each group. For example, all the metaphors to do with movement, progress, motion or trip are placed in one group and given a journey title. I borrowed these metaphor categories from some linguistic studies on environmental problems, like Semino, 2005; Charteris-Black, 2014, 2018; Fløttum, 2017. After that, I conducted a quantitative analysis where I counted the metaphorical expressions in each group and found the proportion of the frequency of each group out of the total metaphors in the four groups

in order to find out what the most common metaphors that are used by the politicians. Besides, at this stage, following Charteris-Black, 2018, I used Lakoff's cross-domain mapping (the source and target domains of metaphors) to offer a deeper analysis of how metaphors were used by the politicians to conceptualize climate change and how the politics of climate change is represented. I also analyzed the conceptual metaphors using a cognitive framework, which is *the metaphor opposition* . In this book, however, it is used to make a distinction between favored (positive) and unfavored (negative) elements of a conceptual metaphor (e.g., journey) for the purpose of explaining how the addressee can be directed (through metaphors) to perceive realities and act accordingly from a cognitive perspective. In this regard, it can be argued that the concepts making up the favored part of the reality encourage the addressee to adopt what is called for and the unfavored (oppositional) concepts, in turn, serve as a warning, pushing him/her to adopt the reality and act accordingly. Consequently, favored and unfavored aspects both push the addressee to comply with the politician's argument. I also combined the two theories (mapping and metaphor opposition) through 'feeding'. This term 'feeding' is used in phonology to mean "to open the door for the other rule (sound change) to apply". Accordingly, in my work, I argue that mapping feeds metaphor opposition. That is, the addressee cannot infer a consequence unless he/she successfully perceives metaphor. In turn, he/ she cannot act accordingly unless this inference is made on the part of the addressee. This book thus expands critical metaphor analysis and critical discourse analysis with this cognitive analysis for the purpose of offering deeper explanations of not only how the addressee perceives realities but also how he/she can comply with the speaker's desires. That is, how the addressee may respond to the speaker's requests (e.g., to fight climate change due to its disastrous consequences) since the speaker implicates a warning of consequences or a motivation (perceived) through metaphor. Thus, mapping and metaphor opposition are developed to explain this from a cognitive perspective.

4. Metaphor explanation: This stage offers explanations of why the speaker/the writer uses certain metaphors (e.g., Journey and/or War

metaphor) rather than others. That is, "the choice of one metaphor rather than another has consequence for how a particular issue is framed or structured, which aspects are foregrounded and which backgrounded, what inferences are facilitated" (Semino, 2008, p. 91). This level also involves determining the purposes for using particular metaphors (e.g., war), and thereby uncovering underlying ideologies and political realities/myths, according to Charteris-Black, 2018. I explained how features of language lead to social change. Change that is socially constructed that the speakers try to make (e.g., hegemonic attitudes and ideologies that they try to form and circulate as facts). At this level, this book is concerned with what motives pushed these politicians to produce these texts. Thus, at this level. I also explained what social changes that the participants try to make including sociocultural views about climate change such as what and how measures should be taken to tackle the problem. Metaphors, which constitute an important path to forming the social world, are the speaker's attempts to persuade the addressee. These metaphors that the participants (the governors & mayors) used are regarded as discursive practices that connect with, and lead to, sociocultural views through which they formulate and justify their arguments. These views represent the goals that politicians aim to achieve so that they can affect the addressee's thinking of the social world (e.g., how they should think of climate change and how it should be addressed?).

Table 4.1 below summarizes how the data is analyzed using the analytical frameworks adopted in this book (Charteris-Black's *critical metaphor analysis* and Lakoff and Johnson's *cross-mapping domain*). For each kind of the conceptual metaphors, I went through these stages.

Metaphors analyzed from many politicians at the federal, state and local levels. These politicians are senators, governors, individuals and among others either supported or opposed the federal government's decision to withdraw from the Paris climate agreement in 2017. The governors and mayors who opposed the federal government's decision in the U.S. These governors, who are seventeen, issued one statement each and a letter by 12 governors. Also, there are 16 senators who supported the U.S. withdrawal decision. In addition, another statement issued by 407

Table 4.1 The stages in the critical metaphor analysis

	Stages in critical discourse analysis
1	Contextual stage: indicating linguistic and social context (situational and cognitive circumstances). This includes investigating why and how speakers/writers produced speeches/texts.
2	Identification stage: linguistic analysis Linguistic representations or surface realizations of climate change (metaphorical expressions) are identified.
3	Interpretation stage: discursive analysis. The metaphorical expressions are placed in groups based on the source domain and from these groups, conceptual metaphors are realized. Source-target mappings and metaphor opposition s are used to explain how inferences are made.
4	Explanation stage: social analysis Offering explanations of what social changes (e.g., attitudes need to be adopted, reforms called for...) that the linguistic features trigger. How is the problem defined and how it should be tackled?

and a letter by 63 mayors in 2017 when the President expressed his intention to withdraw from the United States. Also, this book is only limited to the period between March 2017 to December 2017. It was the time when the plan was announced, and the decision was made by the federal government to withdraw from the Paris agreement. This sample collected from different websites containing this data.

Furthermore, although there are many linguistic and rhetorical features in the data collected, this book only investigates metaphors, as ideologically loaded linguistic features that are discursively used by the politicians. It highlights what metaphors are used and explains why these metaphors in particular are used. These features have a role in not only constructing perspectives towards the social world but also in reformulating stances and perspectives so that social change can be triggered, and thereby the politics of climate change is conceptualized through such features in the U.S.

For the corpus analysis, this book is limited to investigate media texts that are available in the NOW corpus in one year/one month before and after the announcement in both U.S and world media using Collocate, Compare and Frequency tools available in the NOW corpus program.

This book thus draws on qualitative and quantitative methods to enhance the validity of the results. I follow steps-based methodology in identifying and grouping metaphors as well as classifying metaphors into conceptual metaphors. Accordingly, inferring conceptual metaphors is based on concrete steps (based on the semantic content of the metaphorical expressions) in an attempt to avoid inaccuracy as much as possible.

I also used statistics (the frequency and proportions) in the metaphor analysis chapter (the qualitative chapter) to find the frequency and percentage of metaphorical expressions of each type of the conceptual metaphors. Accordingly, the findings of this book are not only based on interpretations and explanations but also on statistics as one way to minimize or avoid subjectivity, and thereby make the findings more valid. In addition, as this book about the debate regarding the announcement of withdrawing from the Paris climate deal in 2017. I included a chapter in this book where I used corpus analysis to search for the most frequent items stringed with the climate change term in the U.S. and world media to see how climate is framed in the media. This is one way to see whether politicians framed climate change in a similar way to that of the media in the U.S. and world media. Additionally, I used corpus analysis to look at the differences in terms of the frequency of occurrences of climate change terms and Paris agreement terms before and after the announcement. In this analysis I applied statistical tests: Log-likelihood and Log Ratio to see whether these differences are statistically significant in the U.S. and World media. This can show to what extent this announcement sparked a heated debate (between the parties) in the world and in the U.S. In conclusion, throughout the analysis in this book, I tried to follow concrete steps, where possible, that I can rely on to investigate how climate change is conceptualized and framed.

Strategies for Sampling

This book uses the qualitative analysis to explore how climate change is conceptualized in the language of politicians in the U.S. The data of the qualitative analysis is the statements issued by seventeen governors and a letter by 12 governors as well as a statement by 407 mayors and also a letter by 63 mayors (when the President decided to pull out of the Paris agreement in 2017). Fifteen senators each issued statements, which supported or opposed the withdrawal. Also, a letter signed by 16 senators supported the decision. Although they are short, they are rich in many linguistic features especially metaphors. Additionally, this book includes a corpus analysis chapter to explore how climate change is framed in the media discourse and if the announcement of withdrawal from the Paris agreement sparked heated debate in the media (through using corpus tools).

For the data of the corpus analysis of this book, it comprises all the news texts available in the NOW program in specific periods (News on the Web). The NOW corpus involves web-based newspapers and magazines from 2010 till this day.

As the main focus of this book is on how climate change is framed in the language of politicians at the executive and legislative level (through their responding to the President's decision of the U.S. withdrawal from the Paris agreement), this section explains in detail how the politicians' data (statements and letters) was collected and why it was studied.

This book investigates texts produced by the President, senators, governors and mayors who represent the executive and legislative structure at the federal, state and local level respectively in the United States. For the statements and letters written by the senators, governors and mayors , they were chosen as the politicians produced these texts formed an alliance against the President's decision at the executive level. These statements and the letters are publicly available. They were collected from the governors' formal websites and the websites of their states and the 'climatemayors' website. Besides, they were published in many websites including social media platforms, including Twitter. In these statements and letters, they (the senators, governors and mayors) expressed their opposition to the President's decision, declaring their stances and views on the Paris climate agreement. In fact, they reflect an important part of the heated debate (between republicans and democrats) over climate change policies in the U.S. These statements and letters, by the governors and mayors, which are the sample of this book, involve a lot of metaphorical expressions drawn from different everyday experiences (e.g., journey and construction). By these metaphors, the governors and mayors opposed, and justified their opposition to, the decision of withdrawing the U.S. from the Paris agreement and shaped realities about climate change as well. Accordingly, the data (the statements and letters) of this book can answer the research questions (as this data involves a lot of metaphors, which can be analyzed to investigate how climate change is conceptualized through metaphor in the U.S). Besides, and in accordance with this conceptualization, this book investigates what the socio-cultural views that are formed and that can be explained from texts examined. I investigated the politics of climate change in social and political contexts in which texts are produced, showing how climate change is conceptualized in the language of politicians.

Table 4.2 shows the number of senators, governors and mayors who reacted to the President's announcement (of withdrawing the U.S. from the Paris climate agreement) by issuing statements and publishing them publicly and also writing letters to the President. Besides, this Table 4.2 gives information of how many texts were produced by these politicians (the governors and mayors).

The politicians reacting to this decision represent two levels of government: the state level represented by the governors and senators and the local level by the mayors. The senators, governors and mayors are the social actors (the participants), who produced those texts (the data), in the book. The following sections provide background information about the senators, governors and mayors .

The Governors

Table 4.3 shows a list of the governors who opposed the federal government's decision to abandon the Paris climate agreement. In 2017, each of

Table 4.2 Distribution of senators, governors and mayors who issued statements and wrote letters

Genre	Statements			Letters		
Politicians	Senats	Govs	Mays	Senats	Govs	Mays
Number of politicians	10	17	407	14	12	63
Number of texts	10	17	1	1	1	1

Table 4.3 The Governors opposing the federal government's decision

	State	Governor	The serving periods	Party
1	California	Gov. Jerry Brown	1975 to 1983; 2011 to 2019	Democrat
2	Colorado	Gov. John Hickenlooper	2011 to 2019	Democrat
3	Connecticut	Gov. Dannel Malloy	2011 to 2019	Democrat
4	Delaware	Gov. John Carney	2017	Democrat
5	Hawaii	Gov. David Ige	2014	Democrat
6	Maryland	Gov. Larry Hogan	2015	Republican
7	Massachusetts	Gov. Charlie Baker	2015	Republican
8	Minnesota	Gov. Mark *Dayton*	2011 to 2019	Democrat
9	Montana	Gov. Steve Bullock	2012	Democrat
10	New York	Gov. Andrew Cuomo	2011	Democrat
11	Ohio	Gov. John Kasich	2011 to 2019	Republican
12	Oregon	Gov. Kate Brown	2015 to now	Democrat
13	Pennsylvania	Gov. Tom Wolf	2015	Democrat
14	Rhode Island	Gov. Gina Raimondo	2015 till now	Democrat
15	Vermont	Gov. Phil Scott	2017 till now	Republican
16	Virginia	Gov. Terry McAuliffe	2014 to 2018	Democrat
17	Washington	Gov. Jay Inslee	2012 till now	Democrat

these governors issued a statement in which he/she expressed his/her opposition. In this section, the names of these governors, their states, the serving periods and the party to which each belongs are presented. This information is important as it adds more contextual clues.

As this book investigates the governors' responses (besides to senators and mayors), to the President's decision on the Paris climate agreement, it would be interesting to look closely at why the politicians on the state level can influence climate change policies (in spite of the world nature of the problem). "Climate change is a long-term phenomenon which is world in nature, but experienced locally" (Dahl, 2011, p. 69). In this respect, Konisky (2014) discusses the reasons why states choose to take on a problem that is world in scope:

> First, many states view such policy as contributing to their economic development goals. This is especially true of renewable energy mandates which not only shift electricity generation away from fossil fuels, but also may spur investment in home grown businesses. Second, some states believe that they already begun to experience the adverse effects of climate change,

> such as sea level rise, severe storms, and prolonged droughts. Third, some states adopt policies to project themselves as policy innovators, often with an objective to advance national policy and/or to encourage other states to join their efforts. Last, in many states, policy has been pushed by organized interest groups and advocacy coalitions and/or citizens through direct democracy or litigation (p. 390).

Politicians (both democrats and republicans) urge action on climate change since they believe its consequences. However, the number of Democrats who call for implementing measures to deal with climate change far exceed that of Republicans. Table 4.4 shows that the proportion of Democrats and Republicans out of the seventeen governors opposing the president's decision. It is clearly evident from the Table 4.4 above that most politicians who opposed the decision are Democrats, who support the claim that human activity is the main cause of climate change, and therefore more policies and measures need to be implemented. Simply put, Democrats have been more active on this issue. They constantly argued against the federal government's decision. The survey shown below (Fig. 4.1) indicates that unlike republicans, the majority of democrats believe that human activity is the main reason behind climate change.

Based on the statistics above (Fig. 4.1), it can be argued that disagreement over climate change can be attributed, to a large extent, to political reasons. That is, according to the chart, while the majority of Democrats believe that climate change is human-made and there is solid evidence of climate change, Republicans do not, except for less than fifth. This statistical evidence demonstrates that this problem which was originated in science is now politicized. That is, which party a politician belongs to affects to a large extent which ideology is the politician to adopt.

Table 4.4 The Proportion of Democrats and Republicans opposing the President's decision

Party	Number	Percentage
Democrats	13	76.5%
Republicans	4	23.5%
Total	17	100%

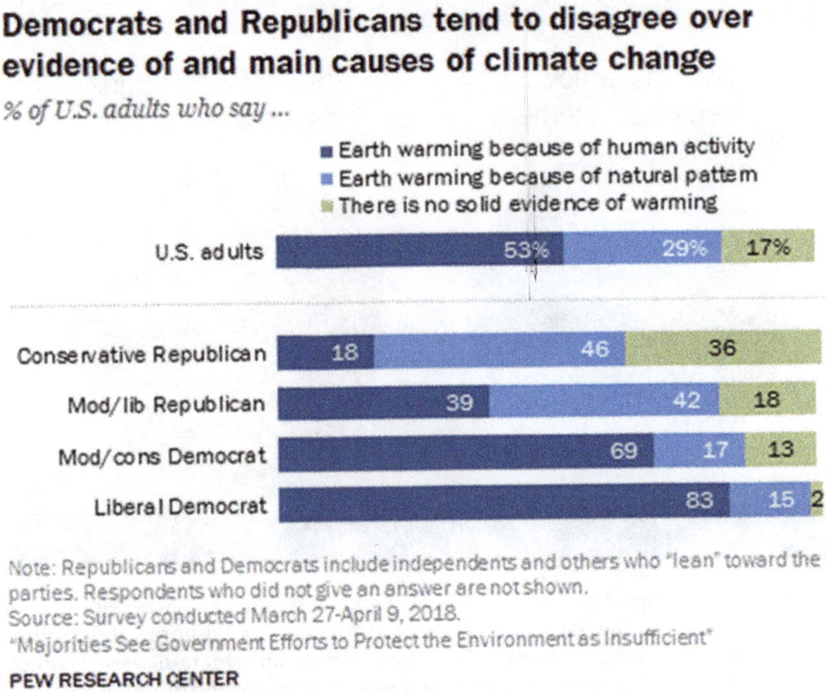

Fig. 4.1 The Distribution of democrats and republicans with regard to their belief in the main causes of climate change. Retrieved from https://www.pewinternet.org/2018/05/14/majorities-see-government-efforts-to-protect-the-environment-as-insufficient/ps-05-10-18_report-10/.

The diagram (Fig. 4.2) below is also another statistical evidence which shows that political party plays an important role in determining politicians' stances towards climate change. The diagram (Fig. 4.2) shows that 76% of Democrats issued statements but 36% of Republicans issued statements opposing the decisions in congress, including representatives and senators.

> There are now two Americas divided along ideological lines- a republican America which questions the validity of climate science and dismisses the urgency of the problem and a democratic America which accepts climate

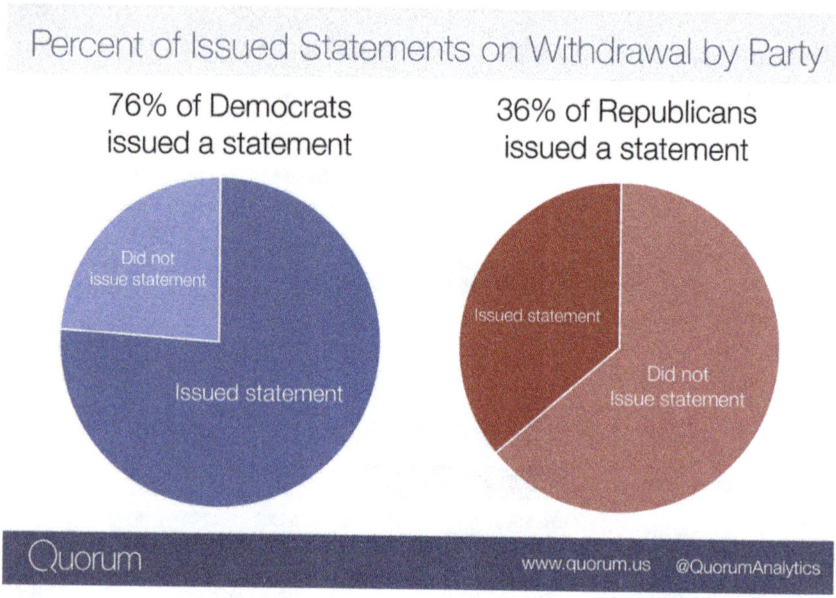

Fig. 4.2 The Proportion of Democrats and Republicans (The House of Representatives and the Senate) who issued statements on withdrawal from the agreement. Retrieved from https://www.quorum.us/data-driven-insights/withdrawal-paris-climate-agreement-reactions-congress-european-union/88/

change science and is concerned about the issue (Nisbet, 2009, cited in Atanasova & Koteyko, 2017, p. 71).

The Fig. 4.2 above shows the percentage of Democrats and Republicans from the House of Representatives and Senate, who reacted against the President's decision on withdrawal of the U.S. from Paris agreement. They issued statements opposing this decision.

> "In the past 48 hours, 184 of the 241 Democrats serving in Congress have responded to President Trump's announcement in an official statement or on social media. However, only 105 of the 291 Republicans serving in Congress have issued responses to the withdrawal" https://www.quorum.us/data-driven-insights/withdrawal-paris-climate-agreement-reactions-congress-european-union/88/.

According to the U.S. Census Bureau, the country includes four government regions. Table 4.5 shows these four regions; and the states which opposed the decision are underlined.

The Fig. 4.3 below shows the proportion of the states opposed the decision in each region. It can be clearly seen that proportion of the states in the Northeast opposing the decision is bigger than that of the other regions, that is 6 out of 9.

For the Northeast states, there are six states represented by their governors reacted against the decision. These states are Connecticut, Massachusetts, Rhode Island, Vermont, New York and Pennsylvania. In contrast, the Midwest has only two states Minnesota and Ohio which responded against the decision. However, in the South, the difference is not that big than the Midwest; only four states out of sixteen states in the South opposed the decision. These states are: Delaware, Maryland, Virginia and California. The West region comes in the second place after the Northeast in opposing the decision, there were six states which reacted against the decision. They are Colorado, Montana, California, Hawaii, Oregon and Washington.

With regards to the governors' environmental activities and achievements I shed light on governors' activities and achievements with regards to protecting the environment in the four official regions, as classified by U.S. Census Bureau: Northeast, Midwest, South and West. This helps explain why these governors in particular stood against the federal

Table 4.5 The four government regions

The official regions	The states in each region
Northeast	Connecticut, Maine, Massachusetts, New Hampshire, Rhode Island Vermont, New Jersey, New York, Pennsylvania.
Midwest	Indiana, Illinois, Michigan, Wisconsin, Iowa, Nebraska, Kansas, North Dakota, Minnesota, South Dakota, Missouri, Ohio.
South	Delaware, District of Columbia, Florida, Georgia, Maryland , North Carolina, South Carolina, Virginia, West Virginia, Alaska, Alabama, Kentucky, Mississippi, Tennessee, Arkansas, Louisiana, Oklahoma, Texas.
West	Arizona, Colorado, Idaho, New Mexico, Montana, Utah, Nevada Wyoming, Alaska , California, Hawaii, Oregon, Washington.

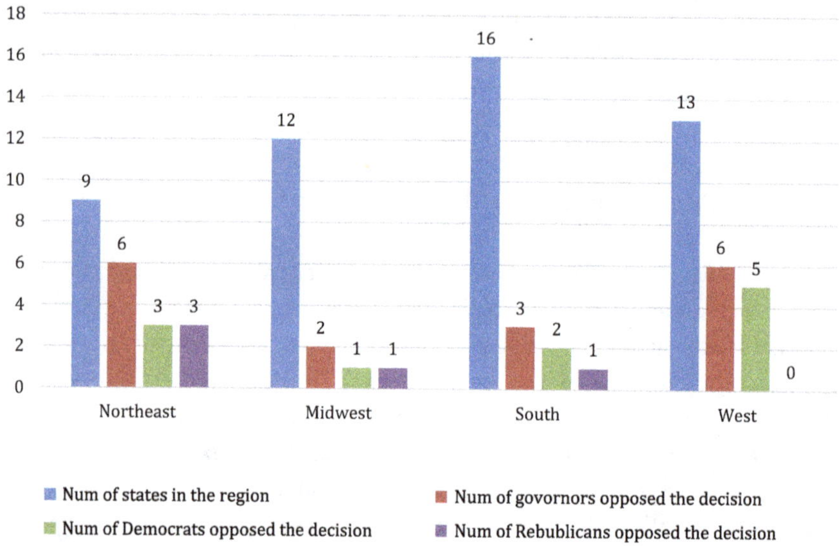

Fig. 4.3 Opposition to the decision by region

government's decision to withdraw the U.S. from the Paris climate accord. For the full description of the activities and achievements of each of the seventeen governors, see appendix C. Since the governors in the Northeast region reacted more than the governors in any other region in the United States, I listed the most important activities and contributions of these governors in the Northeast, as indicated in Table 4.6 (the detailed list of activities for all the seventeen governors are indicated in appendix C). Table 4.6 indicates what the activities that the governors in the Northeast region have achieved with regard to dealing with climate change, according to the national governors association.

Senators

There are many senators who opposed or supported Trump's decision to withdraw from the Paris climate agreement. They used metaphor to draw an image that the Paris agreement has many disastrous or important effects on the economy.

Table 4.6 Environmental activities that the governors in the northeast region have achieved

Northeast	Governor	Environmental activities
Connecticut	Gov. Dannel Malloy	reducing overall emissions; creating good jobs in green economy.
Massachusetts	Gov. Charlie Baker	leading to half of Massachusetts' electricity being generated by clean resources.
Rhode Island	Gov. Gina Raimondo	developing practical solutions to climate change through reductions in greenhouse gas emissions.
Vermont	Gov. Phil Scott	issuing executive order 12-17, forming the Vermont climate action commission. Among the commission's many responsibilities will be the task of unifying Vermont's ambitious climate and economic goals.
New York	Gov. Andrew Cuomo	taking decisive action to lead the fight against climate change and protect the environment for generations to come.
Pennsylvania	Gov. Tom Wolf	supporting and ensuring responsible energy development, while protecting public health and our environment.

"The Senate is composed of senators, each of whom represents a single state in its entirety. Each state is equally represented by two senators who serve staggered terms of six years. There are currently 100 senators representing the 50 states.

WASHINGTON, D.C. – U.S. Senate Majority Leader Mitch McConnell (R-KY) released the following statement regarding President Trump's decision today to withdraw from the Paris climate deal" Wikipedia.

Out of a hundred senators, there are 16 senators who wrote to the president Trump urging him to denounce the Paris agreement. However, there are 10 senators who issued statement opposing Trump's decision to denounce the agreement (Tables 4.7 and 4.8).

The Mayors

While senators and governors represent the legislative and executive branch of government at the state level, mayors represent the executive

Table 4.7 Senators who supported the U.S. withdrawal from the Paris climate agreement

Senator's Name	State
Mitch McConnell	Kentucky
Susan Collins	Maine
Ted Cruz	Texas
Orrin Hatch	Utah
Lindsey Graham	South Carolina

Table 4.8 Senators who opposed the U.S. withdrawal from the Paris climate agreement

Senator's Name	State
Bernie Sanders	Vermont
Tim Kaine	Virginia
Dianne Feistein	California
Kamela Harris	California
Chris Coons	Delaware
Heidi Heitkamp	North Dakota
Ben Cardin	Maryland
Kathleen Corez-Masto	Nevada
Cuck Schumer	New York

branch of government at the local level. There are many duties and tasks that mayors can do. These duties and tasks are basically performed for local people living in cities. In 2017, sixty-three mayors wrote a letter to the President opposing the decision and expressing their wish that the United States remain in the leading position in tackling climate change and remain in the Paris agreement. By writing this letter to the president, they formally opposed the decision (withdrawal from the Paris agreement). The number of mayors opposing the decision steadily increased from sixty-three to four hundred and seven when they issued a statement via their formal website (climatemayors) in 2017. According to the 'climatemayors' website, mayors joined from around the country when they issued this statement, forming an alliance at the local government level against the federal government level. They demonstrated that the ongoing commitment of the U.S. cities to prompt action against climate change was among the main goals of establishing the 'climatemayors' community in 2014:

> Climate Mayors, founded in 2014, is a bipartisan, peer-to-peer network of U.S. mayors working together to demonstrate leadership on climate change through meaningful actions in their communities, and to express and build political will for effective federal and world policy action. The Climate Mayors coalition has emerged as a key voice and demonstration of the ongoing commitment of U.S. cities to accelerate climate progress (http://climatemayors.org/).

The map (Fig. 4.4) shows the cities whose mayors joined the alliance, issuing a statement in response to the decision of withdrawing from the Paris climate agreement. The map demonstrates that the mayors opposing the decision are from cities from all the four regions (Northeast, Midwest, South and West) in the country. Nevertheless, it can be clearly seen that many more mayors who opposed the federal government's decision (to withdraw from the Paris climate agreement) come from the Northeast than the remaining regions, as the map (Fig. 4.4) below shows.

Mayors also have an important impact on policy makers regarding climate change. This explains why mayors opposed the decision to withdraw from the Paris agreement as well. Cities and local governments today are already experiencing the effects of a changing climate and their

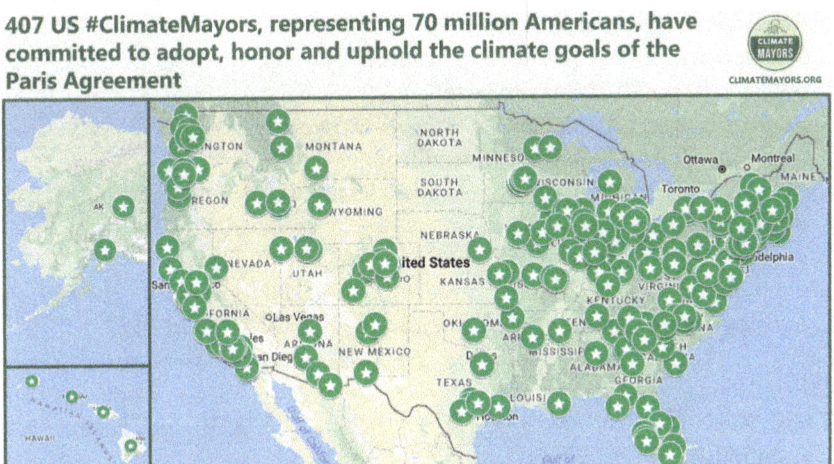

Fig. 4.4 A map illustrating cities which opposed the federal government's decision. Retrieved from the climatemayors website http://climatemayors.org/

exposure and vulnerability to climate-related hazards. The subsequent impacts affecting urban sectors and people in surrounding residential areas is only expected to increase. When addressing climate change, building resilience of urban infrastructure is crucial to guarantee the continuity of services to the population. Cities are not only working on significantly reducing emissions, but are also responding to climate change impacts and preparing themselves for the future. By making adaptation an urban development priority, citizens in GCoM cities have an opportunity for a more resilient future (From: Implementing climate ambition: World Covenant of Mayors 2018 World Aggregation Report).

References

Asplund, T. (2011). Metaphors in Climate Discourse: An Analysis of Swedish Farm Magazines. *JCOM, 10*(04), 1–8.

Atanasova, D., & Koteyko, N. (2017). Metaphors in Online Editorials and Op-ed About Climate Change, 2006-2013: A Study of Germany, the United Kingdom, and the United States. In K. Fløttum (Ed.), *The Role of Language in the Climate Change* (pp. 71–89). Routledge.

Charteris-Black, J. (2005). *Politicians and Rhetoric. The Persuasive Power of Metaphor*. Palgrave Macmillan.

Charteris-Black, J. (2014). *Analyzing Political Speeches: Rhetoric, Discourse and Metaphor* (1st ed.). Palgrave Macmillan.

Charteris-Black, J. (2018). *Analyzing Political Speeches: Rhetoric, Discourse and Metaphor* (2nd ed.). Palgrave Macmillan.

Dahl, T. (2011). Communicating About Climate Change: Linguistic Analysis of Climate Texts. *Journal of Professional Communication, 26*, 69–75.

Fairclough, N. (2014). *Language and Power* (3rd ed.). Routledge.

Fløttum, K. (2017). Willingness of Action. In K. Fløttum (Ed.), *The Role of Language in the Climate Change Debate* (pp. 113–129). Routledge.

Gee, J. P. (2014). *An Introduction to Discourse Analysis: Theory and Method.* Routledge.

Gibbs, W. R. (2017b). *Metaphor Wars: Conceptual Metaphors in Human Life*. Cambridge University Press.

Konisky, D. M. (2014). Environmental Policy. In R. G. Niemi & J. J. Dyck (Eds.), *Guide to State Politics and Policy* (pp. 383–394). Sage.

Pew Research Center: Science &Society. (2018). Retrieved from https://www.pewinternet.org/2018/05/14/majorities-see-government-efforts-to-protect-the-environment-as-insufficient/ps-05-10-18_report-10/.

Quorum. Withdrawal from Paris Agreement spurs reaction from Congress to the European Union. Retrieved from https://www.quorum.us/data-driven-insights/withdrawal-paris-climate-agreement-reactions-congress-european-union/88/.

Semino, E. (2005). The Metaphorical Construction of Complex Domains: The Case of Speech Activity in English. *Metaphor and Symbol, 1*, 35–70.

Semino, E. (2008). *Metaphor in Discourse*. Cambridge University Press.

The climatemayors website http://climatemayors.org/.

Van Dijk, T. (2008). *Discourse and Context: A Sociocognitive Approach*. Cambridge University Press.

Van Dijk, T. (2009). *Society and Discourse: How Social Contexts Influence Text and Talk*. Cambridge University Press.

5

Analyzing Politicians' Metaphors Supporting the U.S. Withdrawal from the Paris agreement

The fifth chapter examines how politicians, who called for and supported the U.S. withdrawal from the Paris climate agreement, used metaphor. I analyzed metaphors (used by the politicians) from CDA theory and Critical Metaphor Analysis perspectives to investigate how climate change is conceptualized in the politics of the United States. At the linguistic level, based on the semantic content, this stage focuses on how metaphors are classified, organized, and arranged. "Metaphors are not inherent in words, but arise from how words are used and understood" (Charteris-Black, 2014, p. 178). This means that source or semantic domains (categories) themselves are not mentioned in the data, but they are realized from the metaphorical expressions used by politicians. I also established another cognitive framework: the metaphor opposition. This framework is used to make a distinction between favored and unfavored elements of a conceptual metaphor (e.g., journey) for the purpose of explaining how the addressee can be directed (through metaphors) to perceive realities and act accordingly from a cognitive perspective. In this regard, I developed this framework to explain how metaphors can serve as a warning or motivation, pushing the addressee to comply with the speaker/writer. Accordingly, conceptual elements making up each conceptual metaphor (e.g., Journey or War) are classified as positive (favored) and negative (unfavored).

O. K. Al-Shboul, *The Politics of Climate Change Metaphors in the U.S. Discourse*,
https://doi.org/10.1007/978-3-031-19016-2_5

Types of Metaphors

The focus of this chapter is on the language of the politicians—the President and the politicians including governors, senators who supported the exit from the Paris agreement, to offer explanations of what the argument or reality that these politicians made. More specifically, I examined the politicians' uses of metaphorical expressions with which they constructed their argument. This chapter includes an analysis of the president's, governors' senators' and mayors' metaphorical choices. This analysis sheds light on linguistic and cognitive aspects of metaphor. That is, I explained why the politicians used such metaphorical expressions and how they used these expressions to shape realities about climate change. This also includes indicating the possibilities of how the addressee can make inferences from these metaphors.

On June 1 2017, the President delivered an official statement announcing that he would withdraw the U.S. from the Paris climate agreement. In that statement, he presented justifications of why he and the administration took this decision. He discursively reconstructed the Paris climate agreement as a disadvantage. He frequently drew on metaphor for this purpose, where he tried to legitimize his decision (withdrawing the U.S. from the Paris climate agreement) that he announced. While many politicians supported this decision, many other opposed it. This chapter thus analyzes the presidents and the politicians who called for the exit as one group or side.

In this speech, the politicians including the President and senators used seven types of metaphors including Impediment, Unfairness, War, Journey, Cleanliness, Construction and Theft, as indicated in Fig. 5.1. Based on the analysis (counting the number of metaphors of each type), Impediment and Unfairness are the most common types of metaphors which were used in their statements and letters as shown in the Table 5.1 below. These two metaphors (Impediment and Unfairness) make up about two thirds of the metaphors used in their statements and letters. This section discusses these seven types of metaphors through examples extracted from the speech, with showing the frequency and percentage of each type of conceptual metaphors used.

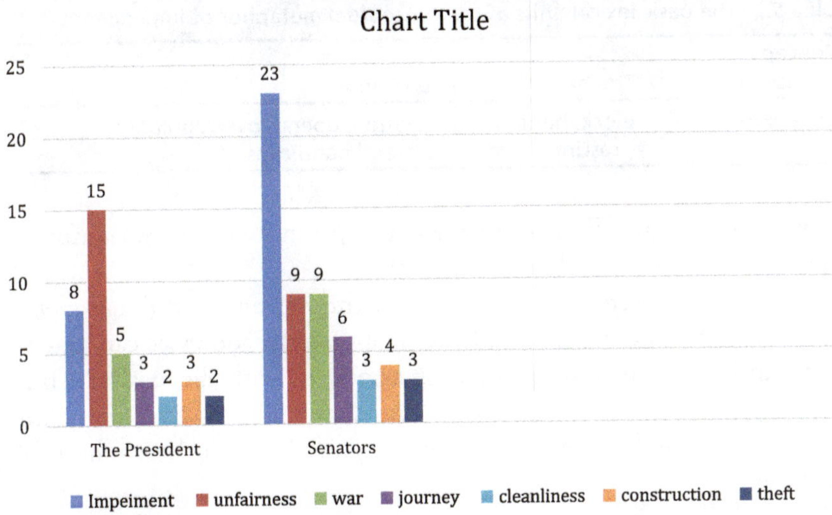

Fig. 5.1 Types of metaphors used

Table 5.1 The types, frequency and percentages of metaphors used by the President and senators

Order	The conceptual metaphor	Frequency	Percentage
1	Impediment	31	32.2
2	Unfairness	24	25.0
3	War	14	14.5
4	Journey	9	9.3
5	Cleanliness	6	6.2
6	Construction	7	7.2
7	Theft	5	5.2
	Total	96	100%

Impediment Metaphor

This is the first most frequent metaphor in this data. There are twenty eight metaphorical expressions which belong to this conceptual metaphor. It accounts for 32.2% of the total (Table 5.1).

The politicians use many metaphorical expressions (as indicated in Table 5.2) drawn from this domain, which is impediment. Those expressions convey the meaning of blockage. They use such metaphorical expressions to construct certain realties about the Paris

Table 5.2 The basic lexical units of the conceptual metaphor of impediment

Conceptual metaphor	The basic lexical units of Journey
Impediment	Block, hindrance , litigation, upend, overregulation, rescind, burden, halt, and handicaps

climate agreement. That is, if we want to promote the U.S. economy, then we must exit the Paris agreement.

The addressee perceives realities about this agreement through metaphors of impediment. Thus, with this mapping the addressee can understand an important implied meaning associated with the axis, which is Paris agreement. Thus, the then President claimed that the Paris Agreement would hinder economic progress by using metaphors of obstacle.

Example 1:

While the current agreement effectively **blocks the development** of **clean** coal in America—which it does, and the mines are starting to open up. We're having a big opening in two weeks. Pennsylvania, Ohio, West Virginia, so many places (The President).

The speaker uses the lexical verb *blocks* as a metaphorical expression of impediment to conceptualize the difficulties and obstacles created by the Paris climate agreement which negatively influence economic developments in the country, such as *clean coal*. Here, the speaker (the President) uses the adjective *clean* from the metaphor of Cleanliness to describe coal conveying the message that coal used is not polluting, yet the Paris agreement limits investing in this industry. He uses the metaphor within a comparative structure (while the main clause, the main clause) to show how the United States are adversely affected due to the Paris climate agreement. He tries to make the addressee realize how the difference is before and after the agreement and also the negative role that the agreement plays with regards to the economy of the country: the agreement blocks coal industry.

The Cognitive Understanding of Impediment as a Conceptual Metaphor

Source-Target Mapping

These concepts making up the source domain of impediment map onto these concepts associated with the target domain (Fig. 5.2). In other words, the A concepts transfer to the B ones. As indicated, all the concepts on the left side are negative, and thereby they map onto negative concepts making up the target domain. Accordingly, obstacles map onto sanctions on the economy; stop progress map onto freezing the economy. This map represents match between the two domains (where these two domains correspond to one another). This inferential mapping is so important for the addressee to process the metaphor and thereby understand what the speaker means. Simply put, the addressee can perceive the reality constructed, which is the Paris climate agreement deters the U.S. economy. This mapping schema triggers another cognitive process embodied in this schema as shown below.

Metaphor Opposition (Impediment)

The process embodied in the image schema (B↔ C) in the Fig. 5.3 above is considered a further cognitive process occurring due to the mapping (A→ B) onto the target domain. Once this mapping happens, opposing concepts to the concepts of the target domain (on the left side) activate.

Thus, the opposition schema (B↔ C) offers an explanation of how the addressee reason about the topic and how the addressee is given a sense of

Source-target mapping

Source domain (A)	→	Target domain (B)
obstacles	→	sanctions on the economy
stop progress	→	freezing the economy

Fig. 5.2 Metaphor mapping from the source to the target domain

Negative (unfavored) concepts (B)		Positive (favored) concepts (C)
sanctions on the economy	↔	facilities to promote economy
freezing the economy	↔	strengthening the economy

Fig. 5.3 The favored and unfavored elements of the entailments

encouragement and motivation to take an action. That part of cognitive thinking then can help arouses emotions. "Metaphors also arouse emotions that can be used as the basis for evaluating political actors and actions by offering persuasive representations of social groups and social issues" (Charteris-Black, 2014, p. 161). If we looked at the concepts (sanctions on the economy, freezing the economy) on the left side in the figure above, we will see that they are negative. These concepts activate opposing concepts (to the ones on the right side). These concepts on the right side are then what pushes the addressee to comply with the speaker's request. That is, if not sanctions on the economy, then facilities can be easily promoted; we should strengthen economy instead of the opposite (which is to freeze the economy). Accordingly, if no freezing for the economy, then strengthening the economy is facilitated. It is thus through this cognitive structure which leads the addressee to perceive this meaning (made by the speaker), thus the perception of the reality made starts with mapping, which in turn enforces reasoning via opposition.

Unfairness Metaphor

There are 24 metaphorical expressions belonging to this type of metaphor. This metaphor makes up 25.0% of the metaphor used in this statement. It is thus the most frequent one (Table 5.3).

Through some metaphorical expressions of unfairness classified under this conceptual metaphor of unfairness, they convey the sense of injustice. The politicians including the president, senators and governors use this kind of metaphor to construct this reality about the Paris climate

Table 5.3 The basic lexical units of the conceptual metaphor of unfairness

Conceptual metaphor	The basic lexical units of Journey
Unfairness	Unfair and fair

agreement. Such metaphorical expressions as unfair belong to the domain of unfairness. They conceptualize the Paris climate agreement as unjust. Using these metaphorical expressions of unfairness with a topic such as Paris agreement generates Semantic inconsistency. This inconsistency elicits the addressee to make inferences as one important way to fix this inconsistency and thereby realize what the speaker means.

Example 2:

For example, under the agreement, China will be able to increase these emissions by a staggering number of years—13. They can do whatever they want for 13 years. Not us. India makes its participation contingent on receiving billions and billions and billions of dollars in foreign aid from developed countries. There are many other examples. **But** the bottom line is that **the Paris Accord is very unfair, at the highest level, to the United States** (President Trump).

The speaker mentions China and India as examples to support his claim and to provide concrete evidence that the Paris accord has double standards. That is, while the Paris accord imposes economic obligations on the United States, it facilitates economic growth in China and India. The speaker uses the comparative analogy to draw the picture of the Paris accord as being unfair. He uses the expression unfair to describe the Paris accord in a clause starting with the conjunction *but* to compare how the United States is treated unfairly against the other countries (China and India). The President depicts the deal as a person with double standards since the deal imposes obligations that are unequal to its members, and the United States is among the countries which suffer from this inequality.

The Cognitive Understanding of Unfairness as a Conceptual Metaphor

Source-Target Mapping

The notions *injustice, dishonesty* and *partiality* representing the meaning of the source domain (unfairness) transfer to the target domain (the Paris agreement). This is because the expressions of unfairness cannot be used with such a topic as Paris climate agreement unless they are taken metaphorically.

In the Fig. 5.4 above, we see the concepts making up the source domain of unfairness map onto the concepts in the target domain. These concepts are negative as they construct the reality that the U.S. has been adversely affected by this agreement. These concepts are injustice, dishonesty and partiality. As these concepts are negative, they map onto matching concepts in the target domain. This mapping from A to B constitutes the addressee's understanding of the topic described by the unfairness metaphor. This understanding via mapping is important as it constitutes the path to another important cognitive process, which is reasoning through which the addressee is motivated to take action on climate change as indicated in the Fig. 5.5 below.

Source domain (A)		Target domain (B)
injustice	→	inequality with other members
dishonesty	→	the agreement items are deceiving
partiality	→	the agreement is biased at the expense of the U.S.

Fig. 5.4 Metaphor mapping from the source to the target domain

Negative (unfavored) concepts (B)		Positive (favored) concepts (C)
inequality with other members	↔	equality with the members
the agreement items are deceiving	↔	agreement should be clear
the agreement is biased at the expense of the U.S.	↔	the agreement should be impartial

Fig. 5.5 The favored and unfavored elements of the entailments

Metaphor Opposition (Unfairness)

In this figure above, the negative concepts trigger further concepts, which are opposite to the concepts on the left side. They (equality with the members, agreement should be clear, the agreement should be impartial) are thus positive. However, they complement the speaker's perception of the topic (the Paris agreement) through the two cognitive structures (starting from mapping to opposition), resulting in understanding and reasoning about the topic on the part of the addressee. They (the concepts which are in the opposing (right) side, as shown in the figure above) represent the positive or favored concepts. Accordingly, activating these concepts in the addressee's mind (as they are the most prominent concepts on the other side associated with the target domain) encourage the addressee to take action, which is what the speaker calls for. They (the activated concepts on the right side) are a motivation, pushing the addressee into further course of action. They can thus arouse the addressee's emotions. This mapping foregrounds the unfairness-associated notions which results in the address's perception of the Paris climate agreement as unfair. It can be apparent that, as shown in the figure above, because these concepts started as negative concepts, opposing concepts (on the right side) which activate are positive, where they encourage the speaker to comply with the speaker's request.

War Metaphor

This type of metaphor is used by those politicians in such new topics (e.g., the Paris climate agreement or climate change in the U.S.) presented to the public as one important way to attract people's attention to what is being addressed. With this metaphor the addressee can perceives the topic as something that has just emerged and needs to be immediately tackled. Additionally, using metaphors of war invite the addressee to think about the seriousness of the topic and the possible deadly or disastrous consequences. This type of metaphor accounts for 14.5% out of the total metaphors. It is the third frequent metaphor used in this statement.

Table 5.4 The basic lexical units of conceptual metaphor of war

Conceptual metaphor	The basic lexical units of war
War	Fight, legal defense, assault, rampage, and hurt

This metaphor is among the most common types. The data analysis shows that there are many lexical items belonging to the journey metaphor used by the politicians (the governors and mayors) in this book. Table 5.4 below shows the different lexical items used to make war metaphors.

Example 3:
I am **fighting** every day for the great people of this country. Therefore, in order to fulfill my solemn duty to **protect** America and its citizens, the United States will withdraw from the Paris Climate Accord.

The speaker used metaphor of War not to urge action on climate change. Instead, he used this type of metaphor to emphasize that he works hard to get a new deal, which empowers rather than undermines the economy of the country, as does the current Paris deal. He used two lexical verbs, which are metaphorical expressions of the conceptual metaphor of war: *fighting and protect.* He used these metaphors to conceptualize his hard efforts that he exercises for the United States. And among these important efforts is his decision to withdraw the United States from the Paris climate agreement. He uses War metaphors to tell the public how much the United States suffers from this accord, and therefore withdrawing from the agreement secures a safer future for the country and its citizens. The speaker tries to make the public realize what dangerous the situation is due to the terms of the Paris climate deal. The speaker announces the decision: *The United States will withdraw from the Paris Climate Accord* after he has used metaphors of War as a sort of justification of why he makes this decision.

Example 4:
"I applaud President Trump and his administration for dealing yet another significant blow to the **Obama Administration's assault on**

domestic energy production and jobs. President Obama made commitments in this deal based off a costly power plan that we knew at the time was on shaky legal ground (The U.S. Senate Majority Leader Mitch McConnell).

The U.S. Senate Majority Leader Mitch McConnell uses metaphor of war. This linguistic expression of war shows to what extent the Obama administration caused damage to domestic fuel energy and jobs as well because of their policies regarding dealing with climate change. Such metaphorical expressions drawn from the war domain convey the atmosphere of seriousness and dangers. This of course makes the addressee think that there should be urgent solutions. The speaker uses such expression to construct certain messages that if we want to protect the U.S., then we must denounce the Paris agreement.

The addressee then perceives realities about this agreement through metaphors of war. Thus, with this mapping the addressee can understand an important implied meaning associated with this axis (Paris climate agreement).

The Cognitive Understanding of War as a Conceptual Metaphor

Source-Target Mapping

With this mapping from the source to the target domain, the meaning deducted is not literal. The addressee is not expected to take the literal meaning when the speaker talks of the Paris agreement in terms of war, as the speaker means the metaphorical war or battle. In other words, *fight* does not cooccur with the Paris agreement or climate change unless it is used metaphorically because on the semantic level, the verb *fight* and *climate change* are inconsistent, but on the metaphorical level they have no clash. Therefore, using such an expression *fight* with an abstract domain such as *climate change* leads the addressee to inferentially transfer concepts that are associated with the source domain of war, from which

Source domain (A)	→	Target domain (B)
trigger	→	prompt action denouncing the agreement.
fighting	→	implement effective measures to develop the economy.
courage	→	ability to denounce the agreement.
victory	→	success to overcome the agreement

Fig. 5.6 Mappings from the source to the target domain

the verb fight is drawn, onto the target domain to solve this semantic inconsistency. This means that implied meaning cannot be processed unless it is taken metaphorically by the addressee. Accordingly, in the Fig. 5.6 above, the concepts of and associated with war: trigger, fighting, courage and victory are transferred onto the domain which needs to be further explained so that the addressee can get the meaning constructed from those concepts from the source domain.

Metaphor Opposition (War)

The figure below shows the conceptual elements that the War metaphor comprises concepts in A as positive elements from one hand; and those existing in B (delay, economy slows down, the agreement maintains its impacts, failure to overcome the agreement as unfavored elements) on the other hand.

The opposing concepts are entailments that are made and activated in the addressee's mind when metaphors of War are used. These concepts represent the unfavored (positive) side of the reality because they are part of, and contribute to, solutions (that they are looking for) to the problem. The other concepts on the right side of the figure (to be accepted, economy slow down, the agreement maintains its impacts) are also entailments which are subsequently made. But they represent the negative or unfavored elements of reality because those concepts make up consequences if the politicians and people could not deal with the problem (exiting the Paris climate agreement). Accordingly, realities made through the conceptual metaphor of War consist of favored and unfavored aspects, where the favored ones encourage the addressee to adopt realities which are shaped using metaphors of War and the unfavored elements serve as a warning, conceptualized by delay, economy slow down, the agreement maintains its impacts for them that they need to deal with the problem (climate change). These elements of both sides are what direct the addressee to specific course of action. Both of which push the addressee to comply with the speaker's request, which is in this context is to support withdrawing from the Paris climate agreement. Accordingly, War metaphor can contribute to shaping the image in the addressee's mind that we are in war. People, therefore, need to realize this reality and act accordingly. This means that the addressee is between two difficult situations that he/she needs to choose: either to refuse the Paris climate agreement or accepts as it is and thereby suffers from those consequences represented by the unfavored elements of the reality made up by War metaphors (in case he/she did not adopt the reality and act accordingly): prompt action denouncing the agreement or to be accepted; implement effective measures to develop the economy or economy slow down; and ability to denounce the agreement or the agreement maintains its impacts and success to overcome the agreement or failure to overcome the agreement. These (favored and unfavored) elements correspondingly contribute to forming the addressee's perception of climate change, determined by this source domain (War). The unfavored concepts then serve as a warning for the addressee. The cognitive structure in the speaker's mind associate negative concepts with negative starting from mapping to opposition. As this conceptual metaphor highlight positive entailments on the left side and thus motivating the addressee, the other concepts of this Fig. 5.7 show how metaphor implies warning as well.

Positive (favored) concepts (A)	↔	Negative (unfavored) concepts (C)
prompt action denouncing the agreement	↔	to be accepted
implement effective measures to develop the economy	↔	economy slow down
ability to denounce the agreement	↔	the agreement maintains its impacts
success to overcome the agreement	↔	failure to overcome the agreement

Fig. 5.7 The favored and unfavored elements of the entailments

Journey Metaphor

As is the case with many topics, climate change can be explained and perceived by means of source domain of journey, in particular. According to many scholars such as Kövecses, 2010 &Semino, 2008, this source domain consists of conceptual elements, which is a set of cognitive schemata: *travelers, progress, the right path and destination*. These conceptual elements (travelers, progress, the right path and destination) make up the pragmatic meaning of journey. They all shape the image of how the speaker wants the addressee to perceive climate change and how it should be dealt with. It is not surprising that the politicians (the governors and mayors) frequently used this type of metaphors. Many studies demonstrate that metaphorical expressions of journey are predominately used in climate change discourse (Charteris-Black, 2005; Fløttum, 2017; Nerlich, 2012; Semino, 2008). Charteris-Black (2005) argues that "the general characteristics of journey [..] make them such an important source of metaphor in political speeches" (p. 207). Atanasova and Koteyko (2017) found that journey metaphors were the most commonly used when they analyzed NYTimes coverage of climate change.

The analysis shows 9 occurrences of Journey metaphors. It makes up 9.3% out of the total metaphors used in the statement. The Table 5.5 below shows different lexical items making up different metaphorical expressions of journey metaphor.

Example 5:
I'm willing to immediately work with Democratic leaders to either negotiate **our way back into Paris**, under the terms that are fair to the United States and its workers, or to negotiate a new deal that protects our country and its taxpayers.

Table 5.5 The basic lexical units of metaphorical expressions of journey

Conceptual metaphor	The basic lexical units of Journey
Journey	Remain, path, continue, maintain, keep joining, remain, future, the way forward, ahead of, right direction, step up, our way back into, and reenter

Here the speaker used noun phrase *our way back into Paris* as a metaphorical expression of journey to ensure that the U.S. is willing to rejoin the Paris climate agreement but under the terms determined by the U.S. He further indicates that the other choice *or to negotiate a new deal that protects our country and its taxpayers*, which is conceptualized by the metaphor of War *protect*. The speaker uses the word fair to identify what kind of agreement that the country is looking for. Also, the use of the adjective fair implies that the current agreement is unfair. Withdrawing from the agreement is, therefore, the only choice left unless there are changes to its terms in ways that do not undermine the economy of the country. He also explicitly stated the agreement is unfair.

Example 6:
Withdrawing from this unenforceable agreement which lacked the support of the American before people and never came before the senate for the approval is **a step in the direction**. Senator Orrin Hatch

The senator (Orrin Hatch) used a metaphorical expression of journey (a step in the direction). In this journey represented by withdrawing from the agreement, they can denounce this agreement which is characterized by the lack of support of the American people. Thus, the agreement is unrealistic and weak that they need to denounce. This withdrawal represents the right step in the desired direction.

Example 7:
I support President Trump's desire to **re-enter the Paris Accord** after the agreement becomes a better deal for America and business (Sen. Lindsey Graham, R-S.C).

Here the senator (Graham) uses an important metaphor. This metaphor implies that the exit is the right thing to do. However, reentry is also something necessary to do but when the agreement becomes a better deal for America and business. The journey metaphor in this example implies two directions. Thus, the exit is something necessary and (but) then the entry is necessary too. Both directions are required as these movements can guarantee a better deal for the U.S. business.

The Cognitive Understanding of Journey as a Conceptual Metaphor

Source-Target Mapping

The Fig. 5.8 below shows the correspondences between the source and target domains shown through mappings between the two domains.

These concepts (travelers, progress, the right path and destination) making up the source domain map onto with the target domain, which is the Paris agreement, including the concepts on the other side (those who negotiate the deal, advances in having better deal, the appropriate items involved in the agreement, and the goal by having a new deal). The speaker uses metaphor of journey constructing certain realities about this agreement. The reality is thus defined in terms of journey domain. These concepts are in terms of the nature of the problem and how it should be dealt. According to many scholars (e.g.,, Musloff, 2016), this mapping occurs due to the fact that journey is semantically inconsistent with the topic (climate change). However, at the metaphorical level there is no clash between journey and climate change. That is, metaphorical expressions of journey are used to conceptualize climate change aspects in terms of journey so that the addressee thinks of climate change in terms of this source domain (journey). With regards to the directionality of transferal

Source domain (A)	→	Target domain (B)
travelers	→	those who negotiate the deal
progress	→	advances in having better deal
the right path	→	the appropriate items involved in the agreement.
destination	→	the goal (getting a new deal for the U.S.)

Fig. 5.8 Mappings from the source to the target domains

or mapping, *travelers, progress, the right path and destination* (of the source domain) are transferred into the climate change domain, and not vice versa. That is, no concepts from the target domain (climate change) are transferred into the source domain of journey. This is because it is journey which is used to conceptualize climate change. Mapping is, therefore, unidirectional (Sullivan, 2017; Wolff & Gentnerb, 2011).

Furthermore, according to Kövecses, 2002; Lakoff & Johnson, 1980; Sullivan, 2017, it is not necessary that all concepts associated with the source domain are activated when it is applied to the target domain. Fun and entertainment are concepts that are associated with the source domain of journey. These concepts, however, are not activated along with the *travelers, progress, the right path and destination* concepts when the journey domain maps onto climate change since climate change is discussed as a serious problem that the world is facing. This means that dealing with this problem is not a matter of fun. This mapping foregrounds the journey-related notions in the address's perception of the Paris agreement and US economy, with the result that the addressee can infer that any effort to denounce the agreement or renegotiate the deal is considered a step or movement forward (progress), and so they should move on and on, following the right direction, to reach the desired destination. Atanasova and Koteyko (2017) argue that journey metaphors on climate change emphasize and show the addressee the right direction to the destination (which is to overcome the problem). Accordingly, all these concepts: *travelers, progress, the right path and destination* make up this conceptualization.

The journey metaphor is being used, the meaning of continuity is constructed and that the path requires rethinking of how they can move through. That is why we find many politicians calling for moving forward and keeping joining the Paris climate agreement for the sake of dealing with this problem. "Journey involves some type of physical movement from a starting point towards an end point" (Charteris-Black, 2005, p. 198); correspondingly, journey metaphor implies kind of path that the politician (the governors and mayors) urges the addressee (policy makers and public) to go through for the purpose of achieving goals and reaching the destination.

Metaphor Opposition (Journey)

When the politicians used the metaphorical expressions of journey, they in fact direct the addressee to certain inferences and entailments to do with journey. That is, when metaphors of journey are used, favored and unfavored concepts associated with journey are activated in the addressee's mind, and both of which contribute to shaping an image about the Paris deal. This image is shaped using the conceptual metaphor of journey.

Based on this Fig. 5.9 above, withdrawing the U.S. from the agreement is a movement forward rather than a movement backward, and consequently the destination may not be reached. The politicians, including presidents, try to associate progress with the new policies that need to be employed, including denouncing the Paris agreement. To understand exactly what the journey metaphors entail and how journey-based entailments constitute realties about the issue, the concepts associated with journey metaphors are activated through *the metaphor opposition*. In this regard, while some aspects are perceived as favored aspects of reality, the other aspects of that reality are perceived as unfavored by the addressee. To put this into practice, when the politicians use journey metaphors, the favored aspects (those who negotiate the deal, advances in having better deal, the appropriate items involved in the agreement and the goal (getting a new deal for the U.S.) associated with the conceptual metaphor of journey are activated in the addressee's mind. These aspects are the ones which the politicians want the addressee to perceive. However, the unfavored aspects (stated on the other side of this figure) are also associated with journey metaphors. They are activated as well. These unfavored aspects are also the ones which the addressee perceive. They serve as a

Positive (favored) concepts (B) ↔	Negative (unfavored) concepts (C)
those who negotiate the deal ↔	no one or very few to negotiate the deal
advances in having better deal↔	current deal weakens the U.S. economy
the appropriate items involved in the agreement ↔	the wrong items in the
the goal (getting a new deal for the U.S.) ↔	same detrimental deal

Fig. 5.9 The favored and unfavored elements of the entailments

warning for the addressee. They (the unfavored aspects) therefore push the addressee to adopt the reality shaped by the politician. These unfavored aspects thus represent the consequences that the addressee may suffer in case he/she does not consider solutions (conceptualized by journey metaphors in the favored side) offered by the politicians. In conclusion, altogether the favored and unfavored aspects contribute to directing the addressee to inferences from which the addressee perceives this reality, which is shaped by using metaphors of journey. While the concepts in the left side in the figure above help the addressee to understand the topic, the negative concepts warn the addressee about the consequences made up by the concepts on the right side (C) so he/she can think seriously of the Paris climate agreement.

Cleanliness Metaphor

This conceptual metaphor only represents 6.2% out of the metaphors used in this statement. The speaker used some metaphorical expressions drawn from Cleanliness metaphor. The lexical item 'clean' was used along with my lexical items of abstract type including break, plan, exit, act. This metaphor denotes purity from any malicious items (Table 5.6).

This conceptual metaphor (cleanliness) indicates that climate change or the Paris climate agreement is identified in terms of cleanliness scope. Such expressions from this domain are predominately used by the politicians for the purpose of encouraging and motivating the addressee to act accordingly.

Example 8:
"That is why it is so important for you to make **a clean exit** from the Paris agreement" (senators' letter to the president)

Table 5.6 The basic lexical units of conceptual metaphor of cleanliness

Conceptual metaphor	The basic lexical units of Journey
Cleanliness	Clean

Here the senators in their letter to the president describe the exit from the agreement as he is cleaning something that is dirty, which needed to be cleaned. This metaphor implies that this agreement seems to involve malicious items. This indicates that if you stay in this agreement, then the agreement will maintain its negative impacts on the country for many years to come. Here the word 'clean' is used to forming the addressee's perception about the Paris climate agreement.

Regarding their discourse "clean exit" there are many instances of this metaphor throughout the letter, which points out to the senator's insistence on leaving this agreement as it has detrimental effects and therefore exit represent an important relief for the country.

Concepts from the source domain highlight so that they can match the target domain. These concepts activated in the mind lie within the border of this target domain (the Paris climate agreement). Thus, for example when someone says if, in reference to the traffic signal that it is red. This means that it is not green and you then have to stop. We think of these two colors in relation to the traffic signal. Thus, if we say red with reference to the signal, then we infer that it is not green. Thus, we say that the signal represents the axis of contrast (between red and green) that these two colors are related in a binary relationship (if it is not red, then it is green; or if it is green, then it is not red). Thus, the activation of one color is in accordance with the other color.

In contrast, if a speaker said that it is not red in general context—without reference to specific context, then it is impossible to infer green as the other color which should be instead, which is on the contrary side. Similarly, if we say that it is unclean with regard to the Paris agreement, then the hearer would infer that the Paris agreement should be clean.

The Cognitive Understanding of Cleanliness as a Conceptual Metaphor

Source-Target Mapping

These concepts (health, clearance, and purity), which make the basic concepts of the source domain of cleanliness, map onto the target domain, forming an important understanding of the topic (the Paris climate agreement). The speaker resorts to the cleanliness metaphor to lead the addressee to certain thinking of the Paris agreement. That is, the Paris agreement harms the U.S. country as it imposes a lot of procedures that they should get rid of them. Thus, this conceptual metaphor comprising these concepts that push the addressee to these inferences that in turn lead him/her for reasoning. Thus, the addressee's understanding of this topic is achieved via mapping (where concepts from the source domain map onto the target domain) as indicated in this image scheme in the Fig. 5.10 above.

Metaphor Opposition (Cleanliness)

In the conceptual metaphor of cleanliness, the opposing concepts (concepts which are on the right side) are activated, as indicated on the right side in the Fig. 5.11 above. They form the addressee's reasoning about the topic, where he/she can draw a consequence. Simply put, the opposition image scheme (B ↔ C) explains this reasoning, which results in warning perception (drawing a consequence on the part of the addressee). Thus, this can trigger the addressee to the further course of action which is to exit or at least oppose this agreement. These concepts thus constitute

Source domain (A)	→	Target domain (B)
health	→	free from malicious items
clearance	→	no more responsibility from the agreement
purity	→	the items are clear

Fig. 5.10 Metaphor mapping from the source to the target domain

Positive (favored) concepts (B)	↔	Negative (unfavored) concepts (C)
free from malicious items	↔	full of malicious items
no more restrictions from the agreement	↔	restrictions on them
the items are clear	↔	items are vague

Fig. 5.11 The favored and unfavored elements of the entailments

warning that the addressee perceives as a result of activating the opposing concepts (C). This warning perceived is what triggers the addressee for showing willingness to comply with the speaker. As shown in the figure, while the concepts (free from malicious items, no more restrictions from the agreement, the items are clear) are positive on the left side, the other concepts are negative or unfavored on the right side (C). On the Right side, they form the warning that the addressee who perceives. This in turn pushes him/her to comply with the politicians: the addressee perceives the consequence in case he/she did not comply.

Construction Metaphor

Just as Cleanliness metaphor, this metaphor represents only 7.2% out of the metaphors used in this statement. There are two metaphorical expressions of Construction metaphor.

The Table 5.7 above lists the basic lexical items which these politicians drew upon. They belong to different parts of speech including verbs, nouns, adjectives and adverbs. This demonstrates that this type of metaphors is widely employed by these politicians.

Example 9:
My job as President is to do everything within my power to give America a level playing field and to **create the economic, regulatory and tax structures** that make America the most prosperous and productive country on Earth, and with the highest standard of living…

The speaker used the verb *create* to conceptualize the foundation that is needed to make the country the most prosperous and productive in the world. The Construction metaphor lexicalized by the clause *to create the*

Table 5.7 The basic lexical units of the conceptual metaphor of construction

Conceptual metaphor	The basic lexical units of construction
Construction	Grow, create, and build

economic, regulatory and tax structures in this example represents the goal which the President aims to achieve; and by using the Construction metaphor, the President tries to motivate and encourage the audience to support his decision regarding withdrawing the U.S. from the Paris climate agreement.

The Cognitive Understanding of Construction as a Conceptual Metaphor

The metaphors drawn from the domain of Construction conceptualize (aspects of) dealing with climate change in terms of the source domain of Construction.

Semantic inconsistency serves as an important factor to elicit the addressee to make inferences. That is, the expressions of Construction cannot be used with climate change unless they are taken metaphorically. Consequently, the notions *integrity, cooperation* and *strength* representing the meaning of the source domain (Construction) transfer to the target domain (climate change). This (inconsistency) is what elicits the addressee to make such inferences and not take the literal meaning, according to many researchers (e.g., Musloff, 2016).

Source-Target Mapping

With this mapping, the addressee understands that the topic, which is the Paris climate change addressed in terms of construction (Fig. 5.12). The constructive aspects direct the addressee into specific inferences, specified by the speaker. Accordingly, this mapping results in shaping the reality that they should have a better deal which can guarantee and maintain rights for the United States. This cannot be achieved unless they all cooperate. This metaphor directs the addressee to the reality that the politicians should work hard for developing the economy. Thus, the

Source domain (A)	→	Target domain (B)
integrity	→	policy makers united by denouncing the deal
cooperation	→	policy makers work collaboratively to negotiate a better deal
strength	→	policies and plans are effective so they can create better economy.

Fig. 5.12 Mappings from the source to the target domain

politicians using this type of metaphors can instill in the addressee the idea that building efforts is necessary. Thus, the concepts on the right side are the further inferences that the speaker makes via mapping. Through these concepts, the speaker understands the topic (through concepts making up the source domain).

Metaphor Opposition (Construction)

This brings us to *the metaphor opposition* to explain how the Construction metaphors can direct the addressee to perceive realities where the addressee perceives the unfavored concepts (as shown on the right side) in this example as warning. This type of metaphor entails that unifying efforts is necessary and important, and there should be more and more cooperation so that diverse or divided efforts, which are the opposite side of the entailments, are avoided. This is because if there were not sufficient cooperation, the basis represented by measures and strategies would be weak and not strong enough to be able to have better deal.

According to the Fig. 5.13 above, the favored concepts stand as opposite to the unfavored concepts which they both, in turn, draw the image of Construction in the addressee's mind. The favored concepts can help encourage the addressee to comply with the politician and the unfavored concepts, in turn, show the necessity of adopting the reality and act accordingly so that they can avoid any bad consequences in case they did not work cooperatively to denounce the agreement; and correspondingly these unfavored elements (dividedness, disunion, weak efforts causing weak economy) can serve as a warning that this procedure, conceptualized by the Construction metaphor, is required if they wanted to avoid dividedness, disunion and weakness, making up the unfavored elements.

Positive (favored) concepts (B) ↔ Negative (unfavored) concepts (C)
policy makers united by denouncing the deal ↔ div idedness
policy makers work collaboratively to negotiate a better deal ↔ disunion
policies and plans are effective so they can create better economy ↔ weak efforts causing weak economy

Fig. 5.13 The favored and unfavored elements of the entailments

This figure above consists of positive or favored concepts on the left side and negative or unfavored concepts on the right side. This demonstrates that this metaphor while it encourages and motivate the addressee to comply with the speaker, it warns him/her of the consequences in case he/she did not comply through all the concepts highlighted. Thus, motivation and warning are what the addressee perceives through this metaphor.

Theft Metaphor

There are only five occurrences of this type of metaphor, with only 5.2% out of the metaphors used in this statement. This metaphor is mainly used to convey message that the economy of other countries is strengthened at the expense of the United States'. Thus, the U.S. economy is weakened because of this agreement (Table 5.8).

Some politicians used metaphor to construct the message that the Paris climate agreement deprived the U.S. rights.

Example 10:
America is $20 trillion in debt. Cash-strapped cities cannot hire enough police officers or fix vital infrastructure. Millions of our citizens are out of work. And yet, under the Paris Accord, billions of dollars that ought to be invested right here in America will be sent to the very countries that have **taken** our factories and our jobs **away from us**. So think of that.

The speaker here resorts to using expressions drawn from the theft/stealing domain to show how the Paris agreement has stolen their rights

Table 5.8 The basic lexical units of the conceptual metaphor of theft

Conceptual metaphor	The basic lexical units of theft
Theft	Away from us, out of, and take a way

to have a better economy. The speaker used the phrasal verb *taken away* as a metaphorical expression of theft metaphor. With using this metaphor, he states that the economy of the country is greatly undermined because many projects are being invested in other countries rather than in the U.S., with the result that many Americans lost their job. The speaker uses this type of metaphor to make the public realize how the country's economy is harmed due to these actions exercised under the umbrella of the Paris climate agreement, and therefore withdrawing from this agreement is a must.

The Cognitive Understanding of Theft as a Conceptual Metaphor

Source-Target Mapping

The source domain map onto the target domain (Fig. 5.14). These concepts (take, deprive rights, remove precedency) on the left side transfer onto the target domain consisting of these concepts: take US economy interests, lose rights of US affecting economy growth and US with less precedency to other countries. This mapping between the domain leads the addressee to an understanding of this problem in a specified way, in terms of theft conceptual domain. Thus, the speaker frames how the addressee perceives the topic (the Paris agreement). This leads the addressee to think of the agreement on climate change as something which undermines the economy. This is because a lot of investments that are supposed to be for the U.S. go to other countries. The concepts on both sides of the mapping in this figure convey the negative aspects, which is something that the speaker wants the addressee to perceive. Via this mapping, the addressee reaches to this understanding. This in turn is the addressee's path to reason about the reality constructed through

Source domain (A) → Target domain (B)
take → take the U.S economy interests
deprive rights → lose rights of the U.S affecting economy growth
remove precedency → the U.S with less precedency

Fig. 5.14 Metaphor mapping from the source to the target domain

which he/she draw a consequence as explained below from a cognitive perspective.

Metaphor Opposition (Theft)

After having the source domain mapped onto the target domain (through concepts transferring onto the Paris agreement), in the metaphor opposition, concepts activate as a result of mapping from the source to the target domain. Further concepts (on the right side) thus activate as a result of cognitive thinking. These opposing concepts push the addressee to think seriously about the Paris agreement (how it is conceptualized). In this Fig. 5.15 above they construct the positive side that the addressee perceives, as encouragement. Thus, give interest for the U.S. interest, gain more rights, foster precedency shape the inferences made by the addressee. Here metaphor convey negative aspects on the left side: (Take the U.S. economy interests, lose rights of the U.S. affecting economy growth, the U.S. with less precedency to other countries), and the right side (give interest for the U.S. instead of other countries, gain more rights, foster precedency to the U.S) in turn convey positive side. While the left side in the figure above represent the concepts of the target mapping to which the concepts of the source domain mapped, the right side in this figure represents the opposing concepts. They activate as opposing concepts in the addressee's mind, with the result they trigger the addressee to comply with the speaker by encouraging him/her through these positive concepts in the right side. They generate motivation for the addressee to comply with the speaker.

Negative (unfavored) concepts (B) ↔ Positive (favored) concepts (C)
take the U.S. economy interests ↔ give interest for the U.S. instead of other countries
lose rights of the U.S. affecting economy growth ↔ gain more rights
the U.S. with less precedency to other countries ↔ foster precedency to the U.S.

Fig. 5.15 The favored and unfavored elements of the entailments

Thus, the politicians supporting the U.S. withdrawal justify their support based on economic factors. However, many business companies issued statements that assert that the Paris agreement does not adversely economy.

"Amazon continues to support the Paris climate agreement and action on climate change. We believe that robust clean energy and climate policies can support American competitiveness, innovation, and job growth," the Amazon company said.

Apple CEO Tim Cook also said "I spoke with President Trump on Tuesday and tried to persuade him to keep the U.S. in the agreement. But it wasn't enough," Cook said in a memo to Apple employees. "Climate change is real and we all share a responsibility to fight it. I want to reassure you that today's developments will have no impact on Apple's efforts to protect the environment. We power nearly all of our operations with renewable energy, which we believe is an example of something that's good for our planet and makes good business sense as well."

Conclusion

The politicians, supporting the U.S. exit, referred 64 times to the Paris agreement throughout their statements. They used many terms to refer to this agreement, including *the Paris accord, the Paris agreement, the Paris accord, the Paris agreement, the Paris deal, this deal, the deal, this agreement, an international agreement, the current agreement, the Paris framework, these agreements*. However, they referred to climate few times. This is mainly when they claimed that the agreement works for the interest of other countries but not the United States: "This agreement is less about *the climate* and more about other countries gaining a financial advantage over the United States....". Accordingly, and based on the metaphorical analysis of their language, the President used metaphor not to address climate change itself (as being threatening or a top priority or anything else) but they used metaphor to address the Paris climate agreement. With these types of metaphors, he shaped the reality that the Paris climate agreement has double standards by its being fair to the other countries but unfair to the United States and it adversely influences the economy of the country. Simply put, throughout their statements, the politicians used all these types of metaphors to convince the audience that the agreement is working for the interest of other countries but not the United States because obligations and responsibilities are not equally distributed, and therefore they believe that the United States cannot maintain in this agreement.

At the cognitive level, where linguistic and psychological aspects are in focus, metaphor includes mapping and reasoning. While A and B (A → B) represent the inferential mapping where the addressee understands metaphor, B ↔ C represents that side of cognitive thinking through which the addressee perceives metaphor as a warning or encouragement. It is also what leads to triggering emotions. It can thus push the addressee to comply with the speaker's request (e.g., calling for denouncing the Paris agreement) by triggering his/her emotions.

In this figure, it can be claimed that if A is positive and eventually B, then C is negative; and if A is negative and eventually B, then C is positive. However, whether C is positive or negative, it explains how it pushes the addressee to think seriously of the problem by encouraging or warning him or her to comply with the request. For example, a speaker said the Paris agreement is unfair. This tails that the agreement is not and so should be fair.

Thus, the figure below provides an understanding of the opposing concepts activation which is considered an important step. Two colors are brought in my mind when I hear the traffic signal, which represents the dimension where two opposing points stand: red or green. If I said that the signal is green, then it is not red, and vice vera.

The signal if not red, then it is green. And if it is not green, then it is red. The signal or the car is as climate change or Paris agreement. In this context, politicians either call for dealing with climate change or not; and US should keep joining Paris agreement or it should denounce it. These opposing concepts lead the addressee to think of what the case would be. What is activated in the addressee's mind after mapping is the opposing concepts of the target domain. Encouragement or warning is thus perceived through such opposing concepts, with these concepts being associated with encouragement or warning.

References

Atanasova, D., & Koteyko, N. (2017). Metaphors in Online Editorials and Op-ed About Climate Change, 2006–2013: A Study of Germany, the United Kingdom, and the United States. In K. Fløttum (Ed.), *The Role of Language in the Climate Change* (pp. 71–89). Routledge.

Charteris-Black, J. (2005). *Politicians and Rhetoric. The Persuasive Power of Metaphor*. Palgrave Macmillan.

Charteris-Black, J. (2014). *Analyzing Political Speeches: Rhetoric, Discourse and Metaphor* (1st ed.). Palgrave Macmillan.

Fløttum, K. (2017). Willingness of Action. In K. Fløttum (Ed.), *The Role of Language in the Climate Change Debate* (pp. 113–129). Routledge.

Kövecses, Z. (2002). *Metaphor: A Practical Introduction*. Oxford University Press.

Kövecses, Z. (2010). *Metaphor: A Practical Introduction* (2nd ed.). Oxford University Press.

Lakoff, G., & Johnson, M. (1980). *Metaphors We Live By*. The University of Chicago Press.

Musloff, A. (2016). *Political Metaphor Analysis: Discourse and Scenarios*. Bloomsbury Publishing.

Nerlich, B. (2012). Low Carbon Metals, Markers and Metaphors: The Creation of Economic Expectations About Climate Change Mitigation. *Climate Change, 110*, 31–51.

Semino, E. (2008). *Metaphor in Discourse*. Cambridge University Press.

Sullivan, K. (2017). Conceptual Metaphor. In B. Dancygier (Ed.), *The Cambridge Handbook of Cognitive Linguistics* (pp. 385–406). Cambridge University Press.

Wolff, P., & Gentnerb, D. (2011). Structure-Mapping in Metaphor Comprehension. *Cognitive Science, 35*, 1456–1488.

6

Analyzing Politicians' Metaphors Opposing the U.S. Withdrawal from the Paris Agreement

In the sixth chapter, I analyzed the counterargument (politicians arguing against the U.S. withdrawal from the Paris climate agreement), I analyzed metaphors (used by the senators, governors and mayors who opposed Trump's withdrawal from the Paris climate agreement employed and those politicians who supported that decision) to investigate how climate change is conceptualized in the politics of the United States. I explained why the politicians who opposed Trump's decision to withdraw the U.S. from the Paris climate agreement used such metaphorical expressions and how the politicians used these expressions to shape realities about climate change. At the linguistic level, based on the semantic content. This stage focuses on how metaphors are classified, organized, and arranged. Similarly, I explained how metaphors work using mapping and metaphor opposition for providing a deeper understanding of how the addressee can be directed (through metaphors) to perceive realities and act accordingly from a cognitive perspective.

Types of Metaphors

This chapter investigates the linguistic and cognitive aspects of individual metaphors used by the senators, governors, and mayors who argued against the U.S. withdrawal from the Paris agreement in their statements

O. K. Al-Shboul, *The Politics of Climate Change Metaphors in the U.S. Discourse*,
https://doi.org/10.1007/978-3-031-19016-2_6

and letters. This chapter thus examines the other (opposing) argument, made by those politicians who opposed the exit.

Similar to what I did in the previous (fifth) chapter, I followed Charteris-Black's framework (2018), but I developed this framework at the cognitive level for the purpose of studying metaphor from a more comprehensive point of view and providing a deeper analysis to metaphor in political contexts.

Before discussing each type of metaphors, as part of linguistic analysis, quantitative information of all four metaphors is presented to indicate the frequency of metaphorical expressions of each type and the frequency of the politicians who responded to the decision in each region in the country with percentages for the purpose of providing more concrete evidence that can corroborate and back up the qualitative analysis of this book.

The Fig. 6.1 indicates the distribution of metaphors used by both groups (the governors and mayors).

As shown in the Fig. 6.1, the governors used four metaphors with different frequencies. The journey metaphor is the most common type used by both groups (senators, governors and mayors). However, while the

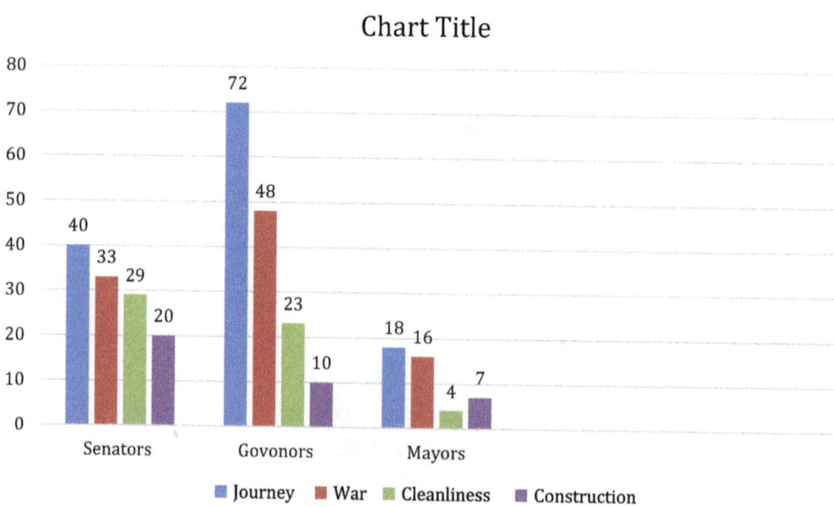

Fig. 6.1 The Frequency of metaphors by the senators, governors and mayors

Table 6.1 The total percentage of metaphors used by both the governors and mayors

Order	Conceptual metaphors	Governors' metaphors Total	Senators' metaphors Total	Mayors' metaphors Total	Total	Percentage %
1	Journey	72	40	18	130	38.5%
2	War	48	33	16	97	28.7%
3	Cleanliness	23	29	4	56	16.6%
4	Construction	10	20	7	54	16.0%
	Total	153	122	45	337	100%

Construction metaphor is the least common metaphor in the sample of the governors' metaphors, the Cleanliness metaphor is the least common in the sample of the mayors' metaphors.

Table 6.1 shows the total number of metaphors used by both the governors and mayors for each type of the conceptual metaphor in this book. The Table 6.1 also gives information about the percentage of each of the four metaphor types. In Table 6.1, the conceptual metaphors are ordered, based on the frequency of metaphorical expressions that each conceptual metaphor has. In the data four types of metaphor are realized: Journey, War, Cleanliness, and Construction. The Table 6.1 also shows that the metaphor that is widely used among the politicians in this book is journey and then war.

Journey Metaphor

In this book, this kind of metaphor is the most common one in this book. It accounts for 38.5% out of the total use of metaphors in the data of this book, making up about the half of the data. The data analysis shows that there are many lexical items which constitute the core components of the metaphorical expressions of the journey metaphor used by the politicians (the governors and mayors) in this book. Table 6.2 shows these lexical units (components) used to make journey metaphors.

When the politician uses metaphorical expressions of journey, the addressee thinks of climate change and dealing with the problem as a journey which in turn direct him/her to identify aspects of reality related

Table 6.2 The Basic lexical units of metaphorical expressions of journey

Conceptual metaphor	The basic lexical units of Journey
Journey	continue, maintain, keep joining, remain, future, staying, the way forward, and path

to movement and destination. This leads the addressee to perceive how movement should be (e.g., forward not backward) to reach the destination, since journey implies advancement and forwardness to reach the goal. The politicians heavily draw on metaphorical expressions of journey since climate change is not a finished process. Change entails something that is still evolving rather than completed (Atanasova & Koteyko, 2017; Cohen, 2011). Consequently, it requires constant and progressive efforts. Additionally, the nature of the problem is complex, and it cannot be easily solved. Therefore, focusing on following steps to deal with such a problem is very important. This explains why politicians used expressions to do with journey a lot. With this type of metaphor being used, tackling climate change is, as Semino (2008) puts it, "conventionally constructed as a destination to be reached, so that movement forwards tend to correspond to positive change, development and success" (p. 81). Another reason why metaphorical expressions of journey are very common and widely used by the politicians is "the availability of a clear schema that includes required elements- such as start and end points connected by a path and entities that move along the path" (Charteris-Black, 2005, p. 46). This makes conceiving climate change as journey is not something difficult to process, and consequently the addressee can understand what the speaker/writer means.

Example 1

America's economy is boosted by **following** the Paris Agreement. From the Governor of California's statement.

The governor uses the verb *following* to show the path to which the U.S. administration needs to go through as an important way to develop the country's economy. The path starts with maintaining joining the Paris climate agreement. Here, the journey metaphor which is lexicalized by the gerund verb *following* has the meaning of encouragement. It carries the message that if we want the economy of our country to prosper, then we need to keep joining the Paris climate agreement. The governor believes that this is a movement forward. He urges people and the government to keep committed to the Paris agreement, which represents an important step that is needed. In this excerpt, the governor uses the verb *boosted.* This verb itself carries a positive connotation where the speaker directs the addressee's attention to the way economy can be rapidly developed; which is by *following the Paris agreement.* By connecting the verb *following* with achieving prosperity (in terms of economy), this gives the verb *following* a sense of an advancement into better. As a result, the metaphor used by the governor demonstrates that climate change is conceptualized as a trip whose goals are strengthening the economy.

Example 2

With or without the support of Trump and the fossil fuel industry we must transition rapidly away from fossil fuels to renewable energy. Sen. Bernie Sanders.

Here the speaker asserts the U.S. transition to the renewable energy which should be done very quickly. He uses the journey metaphor to ensure that the Paris agreement is a step in the right direction to reach the desired destination (renewable energy). This can help the U.S. to leave the fossil fuels for ever after replacing this new form of energy. The journey metaphor used is very important one as it emphasizes this movement regardless of the difficulties that they would have during the process of this transition. The metaphorical expression used here in this example is a verb (transition). It asserts movement to the renewable energy by doing.

Example 3

I hope states around the country **join** Virginia in showing Washington **the way forward** on this critical issue for our nation and our world. From the Governor of Virginia's statement.

The governor of Virginia Gov. Terry McAuliffe uses the journey metaphor to invite all the states in the country to do what Virginia did. By this way, the states can show the President that withdrawing from the Paris agreement is of concern to all the states. They, therefore, are united on their stance on moving forward to dealing with climate change. The governor uses two metaphorical expressions (join and the way forward). These two expressions belong to the same source domain, which is journey. They, therefore, exist in an extended pattern according to some scholars (e.g., Semino, 2008). They both expressions contribute to shaping one message, which is that the governors need to all to have one voice against the decision and also they want to deliver a message to Washington (where decisions are made) that the withdrawal from the Paris agreement is considered a way backward and not forward. Semino (2008) maintains that movement forward denotes progress while movement backward is associated with lack of success. Thus, according to the governor, the way forward cannot be forward if the U.S. pulled out of the agreement. That is, dealing with climate change is the goal that may not be reached if we moved backward (if we did not keep going on this trip which most countries joined). In this example, the governor uses two expressions of movement to indicate that if the destination is to be reached, then the politicians should go through the path that leads them to reach their destination controlling climate change.

Example 4
We urge you to **change course**, and to **join us**. From the mayors' letter to the President.

Here in this example, there are two lexical items (two main verbs) which serve as metaphors of journey. The extended use of journey metaphors (as being coming from the same source domain) in this example help to shape the mayors' message that it is not only to change the decision but also to be an effective member in dealing with the problem. Both these items ask the President to continue to deal with the problem instead of withdrawing from it. Accordingly, dealing with climate change is a predetermined path through which the countries and politicians want to go through, and therefore deciding not to address the problem, according to many politicians such as the governors and mayors, represented by withdrawing from the Paris agreement, is seen as a wrong path or course, which then needs to be changed. The mayors in their letter try to convince the President to take that path by changing his decision (to withdraw the U.S. from the Paris agreement), so they all aggregate on dealing with the problem. The mayors already began addressing climate change and they implemented projects. Thus, they see themselves that they are in the right direction, and they ask the President to join their course so that they walk through one road to reach their destination, which is represented by addressing the problem. Besides to *change course*, the metaphor is lexicalized by *join* to indicate that the only way that the President to be in the right path is to join those who tackle climate change, and the Paris agreement is the right direction.

The Cognitive Understanding of Journey as a Conceptual Metaphor

Source-Target Mapping (Journey)

Climate change can be explained and understood by means of source domain of journey. According to many researchers (such as Kövecses, 2010; Semino, 2008), this source domain consists of conceptual elements, which is a set of cognitive schemata: *travelers, progress, the right path* and *destination*. These elements (travelers, progress, the right path and destination) convey, make up, the semantic meaning of journey. They all shape the image of how the speaker wants the addressee to perceive climate change and how it should be dealt with.

The Fig. 6.2 shows the correspondences between the source and target domains shown through mappings between the two domains.

The politicians consider withdrawing the U.S. from the agreement is a movement backward rather than a movement forward, and consequently the destination may not be reached. For example, the Governor of Virginia Terry McAuliffe in his statement in 2017 stated that "The United States economy is dependent on leadership in the world, yet the President seems inclined to *sit back* and let other nations pass us by….". Also, the Governor of Oregon Kate Brown in his statement: "climate change poses

Source domain (A)	→	Target domain (B)
travelers	→	policy makers implementing measures to deal with climate change
progress	→	advances in dealing with the problem
the right path	→	the appropriate strategies and plans aimed to effectively deal with the problem
destination	→	the goal (overcoming the problem and minimizing it consequences)

Fig. 6.2 Mappings from the source to the target domains

the greatest threat to Oregon's environment, economy, and way of life. Oregon has a strong tradition of fighting climate change, and we will not *back down*….". The politicians, including the governors and mayors, try to associate progress with the new policies that need to be employed including and starting with the U.S. remaining in the Paris agreement. Many metaphorical expressions from the domain of journey are used by the governors and mayors for conceptualizing the problem and how it needs to be addressed. According to Lakoff and Johnson (1980), at the cognitive level, the elements from the source (journey) domain map onto the target domain (in this case, climate change) in terms of how it should be dealt with. By way of explanation, the source domain involving concepts (travelers, progress, the right path and destination) which intersect with the target domain in terms of the nature of the problem and how it should be dealt with carried over to the target domain as mapping occurs. Semantic inconsistency therefore with the topic (climate change) enforces mapping. However, at the metaphorical level there is no clash between journey and climate change. That is, metaphorical expressions of journey are used to conceptualize climate change aspects in terms of journey so that the addressee thinks of climate change in terms of this source domain (journey). With this mapping, the addressee can infer that any effort to deal with climate change is considered a step or movement forward (progress), and so they should move on and on, following the right direction, to reach the desired destination. Journey metaphors on climate change asserts the right direction being followed to the destination (dealing with climate change).

An important meaning that can be inferred from climate change being conceptualized as journey is that climate change is a long-term problem and cannot be treated within a short period of time. Therefore, more and more strategies and procedures are to be implemented by policy makers.

Metaphor Opposition (Journey)

When the politicians used the metaphorical expressions of journey, they in fact direct the addressee to certain inferences and entailments to do with journey. That is, when metaphors of journey are used, favored and unfavored concepts associated with journey are activated in the addressee's mind, and both of which contribute to shaping an image about climate change. This image is shaped using the conceptual metaphor of journey, in terms of favored and unfavored (Fig. 6.3).

In this regard, while some aspects are perceived as favored aspects of reality, the other aspects of that reality are perceived as unfavored by the addressee. These favored ones are shown on the left side (policy makers implementing measures to deal with climate change, advances in dealing with the problem, the appropriate strategies and plans aimed to effectively deal with the problem, overcoming the problem and minimizing it consequences). The unfavored concepts are: no policy makers dealing with the problem, slowness with dealing, the wrong ones, vagueness. To put this into practice, when the politicians use journey metaphors, the favored aspects associated with the conceptual metaphor of journey are activated in the addressee's mind. These aspects are the ones which the

Positive (favored) concepts (B)	↔	Negative (unfavored) concepts (C)
policy makers implementing measures to deal with climate change	↔	no policy makers dealing with the problem
advances in dealing with the problem	↔	slowness with dealing
the appropriate strategies and plans aimed to effectively deal with the problem	↔	the wrong ones
overcoming the problem and minimizing it consequences (the goal)	↔	vagueness

Fig. 6.3 The favored and unfavored elements of the entailments

politicians want the addressee to perceive. However, the unfavored aspects (on the right side) are also associated with journey metaphors. They are activated as well. These unfavored aspects are also the ones which the addressee perceives. They (the unfavored aspects) push the addressee to adopt the reality shaped by the politician. These unfavored aspects thus represent the consequences that the addressee may suffer in case he/she does not consider solutions (conceptualized by journey metaphors) offered by the politicians. While concepts on the left side are positive, the concepts on the right side are negative. Those concepts on the right side make up the warning that the addressee perceives through the inference, and hence they push him/her to comply with the speaker's request. In conclusion, altogether the favored and unfavored aspects contribute to directing the addressee to inferences from which the addressee perceives this reality shaped using metaphors of journey.

War Metaphor

In the book, the data analysis shows that the metaphorical expressions of War accounts for 28.7% of the metaphors used by the politicians, including the governors and mayors. The Table 6.3 on the next page shows the basic lexical units which are used by these politicians in this book that account for the conceptual metaphor of War.

These lexical units belong to different parts of speech including verbs, nouns, adjectives and adverbs. This demonstrates that this type of metaphors (War) is widely employed by these politicians.

This conceptual metaphor (war) indicates that climate change is identified in terms of war scope. Analysis of the data shows that there are

Table 6.3 The basic lexical units of metaphorical expressions of war

Conceptual metaphor	The basic lexical units of War
War	resist, safer, cower, devastate, safer, a fight against, threats, combat, aggressively, damage, deter, dangerous, face, threaten, at risk, the front lines, devastating, protect, secure, and confront.

many metaphorical expressions belonging to this conceptual metaphor *climate change is war*. Such expressions (of war) are predominately used for the sake of prompting action (Cohen, 2011). The politician draws on them to urge the addressee to take an action to confront the problem or otherwise it would be difficult to win. Dryzek (2013) asserts that War metaphor is among the key metaphors employed in the environmental discourse. This type of metaphor is heavily used to conceptualize American politics on climate change. The data analysis indicates that it is the second most prominent metaphor used by the participants. There are two possible reasons for this prominence. The first reason is that this metaphor can invoke within the addressee a sense of urgency, (which is the reason why the politicians issued the statements and letters). That is, people associate war with urgency (something which needs to be addressed immediately, and it should not be postponed), as argued by many researchers (Atanasova & Koteyko, 2017; Fløttum, 2017). The second reason can be that the politicians (the governors and mayors) use these metaphors to invite the addressee to take the issue seriously (Fløttum, 2017) since they believe that climate change poses many threats to the country, and therefore immediate actions should be taken or the situation will get worse. Fairclough (2014) argues that metaphors of War trigger the public's interest. Accordingly, using these metaphors helps the writer create a strong and effective argument so that they can defend their position and convince people of the seriousness of the problem.

Example 5

Our nation has a proud history of leading by example and we do not **cower** from the hard work of making the world a **safer** place. From the Governor of Connecticut's statement.

The governor of Connecticut (Gov. Dan Malloy) here used two metaphorical expressions *cower* and *safer* from the same source domain (war). Such lexical items are commonly used in real war situations and here are used by the politician to conceptualize the metaphorical battle against climate change. Through the metaphorical expressions used in this example, the governor demonstrates that making the world a safer place requires courage rather than cowardice. That is, the governor implies that withdrawing from the Paris agreement is considered cowardliness at the time the U.S. should confront rather than withdraw from the problem or otherwise the world would be at risk. Thus, the governor here shapes an important reality, which is the Paris agreement represents the hard work in dealing with climate change, which needs to be accomplished for the sake of the world's safety of any dangers due to this problem. Hence, the way to overcome this problem is conceptualized as fighting in the battle to protect what is being fought for, which is the environment and all living beings by making the world a safer place.

Example 6

The Paris Climate Agreement is a landmark international **pact** that represents our greatest opportunity to **secure** long-term U.S. and world economic security **in the face of growing threats** from climate change. From the Governor of Washington's Statement.

These metaphorical expressions *pact*, *secure* and *in the face of growing threats* share the concept of war. This example involves extended metaphors since these three expressions are drawn from the same source domain within a single sentence. They conceptualize how climate change should be tackled. Accordingly, the governor of Washington emphasizes that the states need to have courage to confront the problem by showing commitment to the Paris agreement and to protect those who are affected by the problem. The addressee thus realizes that the Paris agreement can provide them with the economic security needed to deal with climate change so that governments, public sectors and private sectors can have more stability with regards to economic as they deal with the problem. Economic security is significant to Americans and Paris agreement is one way to support economic security according to the governor. Here, War metaphors are used by the governor to foreground the notions of safety and security needed to face the problem.

Example 7
That is why we are also **standing up for** our constituents and all Americans **harmed** by climate change, including those **most vulnerable** among us: coastal residents **confronting** erosion and sea level rise; young and old alike **suffering** from worsening air pollution and **at risk** during heat-waves; mountain residents engulfed by wildfires; farmers struggling at harvest time due to drought; and communities across our nation **challenged** by extreme weather. From the Mayors' letter to the President.

Here the mayors in their letter to the President use metaphorical expressions *standing up for*, *harmed by* and *most vulnerable*, *confronting*, *suffering*, *at risk*, and *challenged*. These three metaphorical expressions are all located in a single sentence, coming from the same source domain. All of which are from the domain of war. These expressions represent three different parts of speech, namely, a verb, past participle verb and adjective. This sentence is followed by few clauses to provide more details about how citizens suffer from the consequences of climate change where they employ more metaphors but from the same domain. All these expressions *standing up for*, *harmed by*, *most vulnerable*, *confronting*, *suffering*, *at risk*, and *challenged*, which are extended metaphors (as they draw from one source domain), are used to conceptualize climate change as being threatening showing to what extent people who have been harmed by climate change suffer from many consequences associated with this problem. As a result, mayors need to help and stand with all of those who are suffering from such consequences. This can give a message to the recipient that these all policy makers and governments need to work hard so they can avoid or at least minimize such consequences. One possible interpretation of why the mayors draw on these (war) expressions is to make the addressee realize the seriousness of the situation and this problem needs to be addressed and any delay might cause more problems. The source domain conceptualizes climate change as a real threat that needs to be promptly responded to, or otherwise the situation is getting worse.

The Cognitive Understanding of War as a Conceptual Metaphor

Source-Target Mapping (War)

With War metaphors being used, the addressee would be elicited to realize the seriousness of the problem, and accordingly serious measures need to be implemented. The war metaphor asserts the initial step that needs to be implemented, which is to start dealing with climate change so that they can avoid consequences. Thus, the initial perception that is generated when metaphor of war is used is to focus on the necessity of starting to deal with climate change. This indicates the seriousness and the danger of the situation because of the delay of dealing with this problem.

The concepts *trigger*, *fighting*, *courage* and *victory* correspond into notions in the target domain which make up the semantic meaning of the source domain of war. That is, these concepts are the semantic features (of war) which the war domain involves. Upon the use of War metaphors to talk about climate change on the part of the politicians, the addressee is invited to inferentially process the conceptual metaphor that climate change is war via mapping from the source domain (represented by the metaphorical expressions of war) onto the target domain (Fig. 6.4).

The Politicians using this metaphor in their letters and statements convey the message that they will fight against climate change, standing against any decision that can form an obstacle to addressing climate change. Hence, everyone should have courage and urgently respond to the problem to achieve victory over climate change or otherwise they will lose the battle against that enemy: If they surrendered, there would be

Source domain (A)	→	Target domain (B)
trigger	→	prompt action on climate change.
fighting	→	implement effective measures to tackle climate change
courage	→	policy makers and people's willingness and ability to respond to the problem
victory	→	success to overcome climate change and deal with its consequences

Fig. 6.4 Mappings from the source to the target domain

delay in recovering from the problem, and thereby suffer from serious consequences such as flooding and rise in temperature. Using such metaphorical expressions of war implies transferring concepts that are associated with the source domain of war, onto the target domain. This semantic inconsistency then triggers the inferential meaning implied in the war metaphor. The concepts of war: trigger, fighting, courage and victory are transferred onto the target domain.

Metaphor Opposition (War)

The Fig. 6.5 shows the opposing concepts on both sides A ↔ B, with each side having favored (positive) elements and unfavored (negative) elements. These elements are important for reasoning, which is an important process that the addressee draws upon to reach a conclusion.

Entailments in A that are made and activated in the addressee's mind when metaphorical expressions of War are used. These concepts which are on the left side are the result of mapping from the source to target domain. This Fig. 6.5 shows the positive or favored side (in the left side) and the negative or unfavored concepts (in the right side). Accordingly, realities made through the conceptual metaphor of War consist of positive and negative aspects, where the favored ones encourage the addressee to follow the speaker and the negative concepts serve as a warning, conceptualized by *delay*, *surrender*, *cowardness* and *loss*, for them that they need to deal with the problem (overcoming climate change). The addressee thus perceives the reality in terms of encouragement and

Positive (favored) concepts (B)	↔	Negative (unfavored) concepts (C)
prompt action on climate change	↔	delay
implement effective measures to tackle climate change	↔	measures are weak
policy makers and people's willingness and ability to respond to the problem	↔	no real responding
success to overcome climate change and deal with its consequences	↔	loss

Fig. 6.5 The favored and unfavored elements of the entailments

warning (either to deal with climate change or to suffer from many consequences). These elements then make up the reality which the addressee perceives about this topic. Both of which push the addressee to comply with the speaker's request (which is to deal with climate change).

Cleanliness Metaphor

Cleanliness constitutes an important aspect of people's daily experience. This metaphor is used by those politicians arguing against the exit from the Paris climate agreement to denote any form of energy or any policy that advocates nonpolluting energy (e.g., renewable energy). However, Cleanliness metaphor only accounts for 16.6% of the metaphors used in the data. It comes in the third place in the metaphors in terms of frequency. The lexical units of the metaphorical expressions of Cleanliness are very limited. They are placed in the Table 6.4.

This type of metaphor is mainly represented or realized by the lexical item *clean* in all the data analyzed in this book, and it is accompanied with very limited set of words, which this lexical item (clean) describes: energy, future, plan, power, and technology. Cleanliness metaphors are used by the politicians not only to reflect the policies that need to be followed to deal with climate change but also to influence the decision makers and public in general that the U.S. should use nonpolluting resources (which are clean ones) in order to avoid further consequences that are associated with conventional resources such as pollution. Furthermore, the politicians' use of the lexical item *clean* with energy, power, plan or technology represents a desired goal that the politicians hope to achieve in the country and the world in the near-future. The lexical unit *clean* means in the dictionary free from dirt and uncontaminated, and thereby using this item helps the politician to convince the addressee (people and/or the president) of the necessity of transforming to this type of

Table 6.4 The basic lexical units of metaphorical expressions of cleanliness

Conceptual metaphor	The basic lexical units of Cleanliness
Cleanliness	clean, green, and clear

energy which is described by the word *clean*, which only carries positive meanings.

The lexical item *clean* is semantically used to describe concrete nouns. However, the governors and mayors in their statements and letters use this metaphorical expression *clean* to describe abstract nouns: energy, power, plan or technology, with the result that its conventional meaning (which is free from dirt or stains) transfers onto energy, power, plan or technology. Consequently, the addressee infers that the kind of energy, power, plan or technology which is described by *clean* is nonpolluting and unharming due to the fact that it is conceptualized in terms of the Cleanliness metaphor and, in turn, the other type of energy, power, plan or technology of the current use in the country and the world is not clean or pure (it is polluting), and thereby it has serious consequences, like causing dangerous diseases. The (clean) renewable energy is, therefore, the healthy alternative to that type of (polluting) conventional energy. Thus, the politicians' (the governors and mayors) language of Cleanliness metaphors can push the addressee to think seriously of the causes of climate change, related to the type of current (conventional) energy used. The lexical item *green* is also used by some politicians in the data of this book to mean pure or clean. That is why it was classified as a metaphor under the domain of Cleanliness. It appears with words such as energy and plan to show that new kind of power, which is unpolluting. Li and Ye (2019) demonstrate that "ecological green metaphors in both the East and West are used to advocate environmentalism" (p. 238).

Cleanliness in American culture is an important part of people's everyday experience. It is something necessary if they want to keep healthy. "Cleanliness can be considered a basic domain of experience, due to the human need for being clean in order to preserve health and basic hygiene" (Pavlovic, 2012, p. 29). This type of metaphor can push the addressee to choose and call for employing this type of energy or power described.

Example 8

Hawai'i will continue to fulfill its kuleana on reaching our energy, water, land and other sustainability goals to make island Earth a home for all. The innovation economy is driven by technology, **clean energy**, and **green jobs**. From the Governor of Hawaii's statement.

The governor also uses the word *clean* as an adjective to describe what type of energy that the state is looking for now and in the near future. *Clean* is a metaphorical expression which was used by the governor to describe energy as part of constructing his ideology that any other type of energy is not clean. In addition, the word green (green jobs or green economy) was classified as metaphor under the Cleanliness category since it was used to describe jobs that do not cause any pollution. The governor maintains that one important way to boost the economy is to rely on clean energy and green jobs. "The 21st century has been called the "century of the environment." Governments -and individual citizens—can no longer assume that social challenges such as pollution, dwindling natural resources and climate change can be set aside for future generations" https://fasenet.org/green-technology/.

Example 9

The policies we are implementing that support the U.S.'s achievement of its Paris commitment not only cut carbon pollution—they also create jobs, boost competitiveness, and bring **clean energy** and a cleaner environment to our citizens. These benefits can and should accrue to all Americans. From the Governors' letter to the President.

With this expression *clean energy*, the governors demonstrate that this type of energy as one of the most important goals of the Paris agreement (besides to creating jobs and boosting competitiveness through this agreement). The metaphorical expression *clean* is followed by the noun *energy*. The lexical item *clean* is used to develop a sense of encouragement in the audience regarding what type of energy that is to be used. People are attracted to what it is described as clean and pure. The governors aim to direct the addressee to realize the status of the current energy (which causes pollution) in the state and country.

Example 10

We write to strongly object to your actions to roll back critically important U.S. climate policies including **the Clean Power Plan** and vehicle fuel efficiency standards, as well as proposed budget cuts to the EPA and critical federal programs like Energy Star. From the mayors' letter to the President.

In this example, the lexical item *clean* is used with power plan to describe what type of power plan that the country should have. *The Clean Power Plan* is a noun phrase where the adjective *clean* is associated with two abstract nouns: power and plan. This description is functional. That is, it can trigger important entailments in the addressee; which is the plan for using new types of power recommends using only unpolluting resources. This plan and other standards, therefore, need to be implemented to change the current type of energy.

The Cognitive Understanding of Cleanliness as a Conceptual Metaphor

Source-Target Mapping (Cleanliness)

With this metaphor being used by the politicians, certain aspects linked with Cleanliness are highlighted. Some abstract nouns in the data such as plans, solutions and technology are described as clean, which means unpolluting since they do not produce emissions.

At the cognitive level, due to the semantic clash (which serves as an activator for mapping), the addressee inferentially maps this concrete domain (and the more familiar topic) instantiated by the terms *clean* or *green* onto the target domain: a new and innovative solution (nonpolluting energy) which can be used to deal with and subsequently minimize the risks of the problem. The directionality of mapping as shown in the Fig. 6.6 is one way direction. That is, those concepts: *health*, *freshness*,

Source domain (A)	→	Target domain (B)
health	→	free from diseases that may occur due to climate change
freshness	→	the atmosphere is livable
purity	→	energy consumed is not contaminated

Fig. 6.6 Mappings from the source to target domain

purity making up the source domain transfer into the climate change domain, but not vice versa. Hence, Mapping is unidirectional (Sullivan, 2017).

The Fig. 6.6 shows the notions of the source domain being applied to those of the target domain.

The politicians using the Cleanliness metaphors implicitly invite the addressee to make a comparison between the stuff from his/her everyday experience (when it is clean) with plans, solution and technology being described as clean. Applying the source domain to the target domain, the addressee can infer that energy, plans, solutions and technology, being described as clean objects help him/her avoid many diseases that would be transmitted. As a result, he/she can keep healthy, just like when his/her possessions are clean.

Metaphor Opposition (Cleanliness)

With the metaphor opposition in this conceptual metaphor (Cleanliness metaphor), the meaning or message constructed is that climate change can be addressed by adopting and using a (new) type of energy, which is the opposite of what has been currently used. When politicians use metaphor of Cleanliness (e.g., clean power or technology), this entails that the current type of energy is unclean, and therefore causes harm to the environment. One way to minimize the harm to the environment is to completely shift to clean energy. The current type of energy, however, (unclean) is implicit; it is not explicitly mentioned by the politicians. Clean energy, power, plan or technology represents the choice which the Americans and the world should adopt if they wanted to avoid many consequences (disease, staleness, contamination) associated with the unclean energy.

The metaphor opposition of the Cleanliness metaphor is indicated in the following Fig. 6.7.

Based on this metaphor opposition (and with regards to the Cleanliness domain in describing climate change), reality consists of two sides of aspects: the favored concepts (which are on the left side) and the unfavored concepts (disease, staleness and contamination), which make up

Positive (favored) concepts (B) ↔ Negative (unfavored) concepts (C)
free from diseases that may occur due to climate change ↔ disease
the atmosphere is livable ↔ staleness
energy consumed is not contaminated ↔ contamination

Fig. 6.7 The favored and unfavored elements of the entailments

the Cleanliness domain. Both of which can push the addressee to comply with the speaker and act accordingly.

In sum, it is through these conceptual associations that belong to the Cleanliness domain, the solutions to climate change are conceptualized in terms of Cleanliness. The politicians' message that they try to deliver, through the Cleanliness metaphor, to the audience is conveyed through this conceptualization. That is, it is important that we rely on clean energy, power, plan or technology as alternative resources to the current ones. This keeps us healthy because we can have fresh atmosphere and we are able to avoid many pollution-related diseases since it is ecofriendly and it causes no harm to the environment. This Fig. 6.7 shows positive aspects which motivate and encourage the addressee while the negative aspects of metaphor warn him/her to comply with the speaker's request so that he/she can avoid any bad consequence to the addressee. While the concepts on the left side arise as the result of inferential mapping from (A to B), the concepts (C) on the right side are triggered as opposing concepts to the ones on the left side. This explains why metaphor effectively triggers or moves people to take action. That is, both motivation and warning are important.

Construction Metaphor

This kind of metaphor accounts for 16.0%. It is the least common one in this book. All the lexical units of the metaphorical expressions of the Construction metaphor existing in the data of this book are included in Table 6.5.

Table 6.5 The basic lexical units of metaphorical expressions of construction

Conceptual metaphor	The basic lexical units of Construction
Construction	build, restore, strengthen, create, and coalition.

This kind of metaphor conveys the meaning of building efforts and those efforts should be well built in dealing with climate change:

> Participants also drew on the [construction] metaphor to reason that well-being depends on the work of others. Just as a building is constructed depends on different people with different skills working together with materials, people's well-being is not solely of their own making but depends on what others do and on the resources that are available. The Construction Explanatory Metaphor helps people understand how each and every person's well-being depends on context, resources and the support of others (frameworksinstitute). https://www.frameworksinstitute.org/pubs/mm/talkinghumanservices/page7.html

Cooperation is, therefore, an important factor to succeed. Through the metaphors classified under the conceptual metaphor of Construction, linguistic expressions of Construction metaphor convey cooperation and collected work, like *partnerships, coalition, with our constituents face-to-face, restoring our nation's manufacturing base*. The governors and mayors use this kind of metaphor to emphasize that being unified is a necessary requirement to influence climate change policies. Many of whom have similar visions, but their efforts are still diverse. The efforts thus need to be built and unified so that they can cooperate to deal with the problem. Just like when someone builds something, they need to have workers and materials. The politicians using this metaphor try to conceptualize and explain that cooperation to build and strengthen their efforts is highly important so that they can further implement effective strategies to deal with climate change. This basis represents, and encompasses, politicians' stance on the problem and people's willingness to address climate change and support plans to deal with this problem.

The Cognitive Understanding of Construction as a Conceptual Metaphor

Source-Target Mapping (Construction)

The metaphors drawn from the domain of Construction conceptualize (aspects of) dealing with climate change in terms of the source domain of Construction.

Semantic inconsistency serves as an important factor to elicit the addressee to make inferences. That is, the expressions of Construction cannot be used with climate change unless they are taken metaphorically. Consequently, the notions *integrity*, *cooperation* and *strength* representing the meaning of the source domain (Construction) are transferred to the target domain (climate change). This (inconsistency) is what elicits the addressee to make such inferences and not take the literal meaning, according to many researchers (e.g., Musloff, 2016) (Fig. 6.8).

This mapping results in shaping the reality that dealing with climate change requires cooperation between the policy makers so they can have a foundation upon which they can deal with this problem. In order to do so, as many states and governments as possible need to work altogether so that they can implement more and more effective measures to achieve one goal which is to deal with climate change and minimize its consequences. This cannot be achieved unless they all cooperate. Thus, the governors and mayors using this type of metaphors can instill in the addressee the idea that building efforts is necessary. Simply put, with this conceptual metaphor, the addressee is directed to perceive the reality that all the decision makers at all levels in the states should work hard to be able to effectively respond to the problem.

Source domain (A)	→	Target domain (B)
integrity	→	policy makers undivided by having one goal
cooperation	→	policy makers work collaboratively to deal with the problem
strength	→	policies and plans are effective so they can affect action on climate change.

Fig. 6.8 Mappings from the source to the target domain

Metaphor Opposition (Construction)

This brings us to *the metaphor opposition* to explain how the Construction metaphors can direct the addressee to perceive realities. In fact, this explains why the addressee might perceive warning. It is the role of concepts of the other side that are activated in the addressee's mind. This type of metaphor entails that unifying efforts is necessary and important, and there should be more and more cooperation so that diverse or divided efforts, which are the opposite side of the entailments, are avoided. This is because if there were not sufficient cooperation, the basis represented by measures and strategies would be weak and not strong enough to be able to effectively deal with the problem.

According to the Fig. 6.9, the favored concepts stand as opposite to the unfavored concepts which they both, in turn, draw the image of Construction in the addressee's mind. The favored concepts can help encourage the addressee to comply with the politician and the unfavored concepts, in turn, show the necessity of adopting the reality and act accordingly so that they can avoid any bad consequences in case they did not work cooperatively to deal with climate change; and correspondingly these unfavored elements can serve as a warning that this solution, conceptualized by the Construction metaphor, is required if they wanted to avoid dividedness, disunion and weakness.

Example 11

Vermont has taken a leadership role in addressing climate change, and the President's decision today only **strengthens our commitment**. We have **strong**, regional multi-state **partnerships** through the **Coalition** of

Positive (favored) concepts (B)	↔	Negative (unfavored) concepts (C)
policy makers undivided by having one goal	↔	dividedness
policy makers work collaboratively to deal with the problem	↔	disunion
policies and plans are effective so they can affect action on climate change	↔	weakness

Fig. 6.9 The favored and unfavored elements of the entailments

Northeast Governors (CONEG). From the Governor of Vermont's statement.

The governor of Vermont uses the lexical items *strengthens*, *strong*, *partnerships* and *coalition* to conceptualize the fact that dealing with climate change needs cooperation and steadiness from all the states and that without this, it would be difficult to overcome the problem. These efforts are like blocks that need to be stacked so that they can be strong enough to influence the decision. Thus, here the metaphorical expressions *strengthens*, *strong*, *partnerships* and *coalition* of the source domain can give a message to the addressee that this problem (climate change) is a problem that is not easy to deal with and working altogether can strengthen their stance so that they can overcome such a big problem. The governor of Vermont has used the metaphor of Construction to deliver a message to the audience that the current stance is not solid and well-build enough. Therefore, more work is needed, and this work would be difficult to accomplish without cooperation and partnerships between the states. Accordingly, the metaphorical expressions *strengthens*, *strong*, *partnerships* and *coalition* were used to convey the constructive aspects of how to deal with climate change.

Example 12

As Mayors, we work with our constituents face-to-face, every day, and they demand that we act on climate to improve quality of life and **create economic growth**. From the Mayors' letter to the President.

The mayors demonstrate that dealing with climate change is an important and effective way to improve the quality of people's life and boost their economy. This can be achieved by cooperation and building ties with one another as policy makers. The mayors use the metaphorical expression *create* economic growth to indicate the importance of commitment to cooperate to deal with climate change. This lexical item *create* represents the concrete side and *economy growth* represents the abstract side of mapping, onto which the features of *create* map. The concrete and abstract domains are thus the two contributory sides to one another in conceptualizing climate change in terms of economy prosperity, which can be ensured by and through cooperation.

Example 13

Climate action is also an investment in our economy and job creation—electric vehicles, solar power, energy efficiency and battery storage are all avenues to **restoring** our nation's manufacturing base. ***From the Mayors' letter to the President.***

Mayors conceptualize the attempts to deal with climate change as rebuilding and strengthening the country in terms of the economy. Therefore, the new innovative solutions and measures, including electric vehicles, solar power, energy efficiency and battery storage are all important elements for this rebuilding or restoring. Here the lexical item *restoring* is used to convey some aspects of this construction. In the dictionary, the lexical item restore means to renew and rebuild, and here it is listed under the Construction metaphor as it conveys aspects of renewal and building. The mayors believe that they can treat climate change through relying on *electric vehicles, solar power, energy efficiency* and *battery storage* as a means to build the manufacturing base and create more jobs. Accordingly, these measures seen as a basis upon which they can cooperate at a higher level to deal with climate change.

Based on these source domains, including Journey, War, Cleanliness and Construction identified in the examples above which the governors and mayors used, it is worth discussing how the politicians drew on these domains to construct this issue (climate change) in terms of causes and effects. "Conceptual metaphors reinforce a certain type of perspective, which, in its turn, serves as a scaffolding in cognitive construals of a certain ideology" (Fløttum, 2017, p. 94). The conceptual metaphors that are realized from the linguistic realizations are based on the similarities perceived between some aspects of climate change and some aspects of the source domains from which the metaphorical expressions are drawn (Semino, 2008). Climate change is, therefore, conceptualized based on these similarities of these aspects between the source and target domain.

In journey metaphor, the journey aspects are highlighted and activated. The metaphorical expressions from this type of the conceptual metaphor foregrounds the reality that climate change is not a problem happening over a short period of time. This is due to the fact that climate change is caused by non-end and ongoing factors and in turn the living entities and the environment are suffering from many ongoing consequences. Solutions, which are conceptualized by metaphors as movements forward not backward, therefore, should be of progressive nature. This conceptual metaphor prompts that the solution should be in the long term in form of steps that they need to follow, starting at keeping joining the Paris agreement, to reach their destination, which is to control climate change.

In the conceptual metaphor of War, the addressee is invited to view climate change as an enemy and that people are in a (metaphorical) battle. In this metaphor, aspects of war are activated in the addressee's mind via mapping from the source domain (war) to the abstract domain (climate change) in terms of the nature of the problem and how is to be addressed. Because the problem is life-threatening, solutions should be strict, serious and urgent. War metaphors, therefore, are the linguistic expressions used by the governors and mayors to shape that reality. That is, politicians do not announce war (use metaphors of war) unless there would be something so threatening that they cannot live peacefully unless it is addressed. In this metaphor, the politicians invoke a sense of urgency within the addressee. Thus, the source domain triggers important entailments, including responding to the issue promptly since it represents a real threat to the country and globe since there are people who are harmed by climate change.

In the conceptual metaphor of Cleanliness, the addressee is presented between two types of energy or power used in the country: clean and conventional. Through the inferences that the addressee can make from the Cleanliness metaphors, he/she can realize that the conventional energy is one cause for climate change. At the same time, the Cleanliness metaphors entail solutions accordingly, which is to rely on unpolluting resources, like clean energy and technology.

In the Construction metaphor, the governors and mayors use this type of metaphor since they feel that there is not enough agreement and consensus about one of the most important environmental problems in the world, which is climate change. Hence, they feel that they are standing at a weak base which needs strengthening. Fill (2018), Fløttum (2017) and Hoffman (2012) maintain that there is no social or political consensus on climate change despite the scientific evidence and the scientific consensus on climate change. This is one important reason of why this problem (climate change) is difficult, not to mention that this problem itself is complex. "When human decision systems (be they individuals or collective bodies such as governments) confront environmental problems, they encounter two orders of complexity" (Dryzek, 2013, p. 9). He further mentions the first order of complexity, which is "ecosystems are complex", and the second order is that "our knowledge of them is limited…. Thus doubly complex" (Dryzek, 2013, p. 9). The politicians, therefore, used metaphorical expressions of Construction, which entails the importance of integrity and cooperation, to emphasize that collected work is an important way to tackle this problem to face such difficulties and complexities.

Conclusion

The metaphor at the cognitive level is then based on two image schemas. The first image schema is mapping (where concepts map from the source into target domain). The second image schema is the opposition (thus, concepts exist as binary). These schemas are necessary to explain, from psychological and cognitive perspectives, how the addressee understands the topic through metaphor and how he/she reasons about the meaning and so shows willingness to act accordingly. It is very important to note that these image schemas are conditional, where the second image schema is conditioned on the first schema. This means that the first image schema feeds the second image schema. It opens the door to happen. Correspondingly, opposition is conditioned on Mapping. Once mapping occurs resulting in understanding the target domain, reasoning about realities perceived occurs. That those concepts trigger further concepts that activate in the opposition paradigm. These concepts are associated with the target domains. They constitute the opposing side of the concepts of the target domain. This means that drawing a consequence and motivating the addressee (where the opposing concepts activate) does not occur unless the addressee understand metaphor (where the source concepts map onto the target domain). These concepts triggered are what make the addressee think of the speaker's request seriously, and thus extending the addressee's perception.

Thus, this offers insights into how mapping and opposition work in the mind. The addressee can thus read and understand what the speaker means or implicates. While understanding can occur successfully, it is not necessary that drawing a consequence (reasoning) successfully occurs.

This Fig. 6.10 shows how these cognitive processes occur through the image scheme as presented.

Dimensions of the opposing concepts centered by the topic (climate change or Paris agreement) (Fig. 6.11).

The traffic signal → red ↔ green

If red, it is not green and so it should be green soon.
If not red, it is then green, and so it should be red soon.
If green, it is not red and so it should be red soon.
If not green, it is then red and so it should be green soon.

In fact, these statements above are conclusions that we reach through logical thinking of how the traffic signal works. These conclusions are then reached as a result of activating specific concepts (just as, red or

A → B ↔ C

Fig. 6.10 Image schema

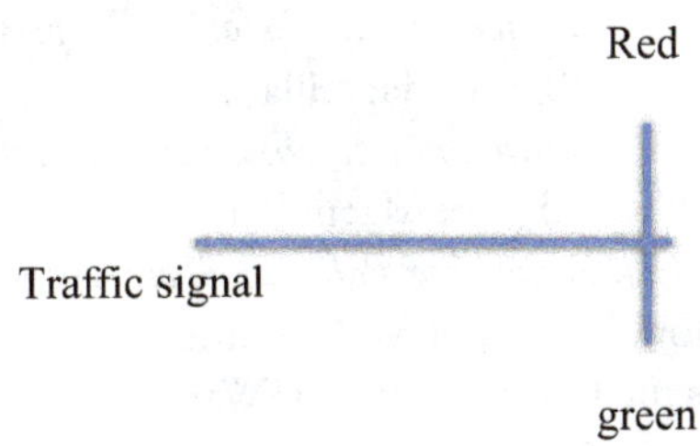

Fig. 6.11 Dimensions of the opposing concepts

green) associated as binary to the axis (just as traffic signal) in this context as an example for clarification.

This is kind of reasoning that the addressee does to draw a conclusion or set of conclusions about the topic addressed. This or these conclusion(s) serve(s) as a motivation or warning to comply with the speaker.

The traffic signal → green ↔ red

The traffic signal → red ↔ green

From a cognitive perspective, this metaphor theory developed in this book demonstrates how metaphor restructures people's thinking of climate change and Paris agreement, and how they may act. The constructive aspects or concepts of metaphors highlighted, as represented by image schemas, have an important role in shaping how the addressee thinks of the problem and acts accordingly. From the source domain, some concepts map, which are most associated with the topic, onto the target domain. This mapping constitutes an important understanding of the target domain. The opposing concepts, however, push the addressee to draw a consequence.

References

Atanasova, D., & Koteyko, N. (2017). Metaphors in Online Editorials and Op-ed About Climate Change, 2006–2013: A Study of Germany, the United Kingdom, and the United States. In K. Fløttum (Ed.), *The Role of Language in the Climate Change* (pp. 71–89). Routledge.

Charteris-Black, J. (2005). *Politicians and rhetoric: The persuasive power of metaphor*. 1st Edition. UK: Palgrave Macmillan.

Charteris-Black, J. (2011). *Politicians and rhetoric: The persuasive power of metaphor*. 2nd Edition. UK: Palgrave Macmillan.

Charteris-Black, J. (2018). *Analyzing Political Speeches: Rhetoric, Discourse and Metaphor*. 2nd edition. UK: Palgrave Macmillan.

Cohen, M. J. (2011). Is the UK Preparing for War? Military Metaphors, Personal Carbon Allowances and Consumption Rationing in Historical Perspectives. *Climate Change, 104*, 199–222.

Dryzek, J. (2013). *The Politics of the Earth: Environmental Discourses* (3rd ed.). Oxford University Press.

Fairclough, N. (2014). *Language and Power* (3rd ed.). Routledge.

Fill, F. A. (2018). Introduction. In A. F. Fill & H. Penz (Eds.), *The Routledge Handbook of Ecolinguistics* (pp. 1–7). Routledge.

Fløttum, K. (2017). Willingness of Action. In K. Fløttum (Ed.), *The Role of Language in the Climate Change Debate* (pp. 113–129). Routledge.

Hoffman, A. (2012). Climate Science as Culture War. *Stanford Social Innovation Review, 10*(4), 30–37.

Kövecses, Z. (2010). *Metaphor: A Practical Introduction* (2nd ed.). Oxford University Press.

Lakoff, G., & Johnson, M. (1980). *Metaphors We Live By*. The University of Chicago Press.

Li, L., & Ye, M. (2019). Greenspeak: A Corpus-based Comparative Study on the Word Green and Environmentalism. *Equinox, 13*, 221–240.

Musloff, A. (2016). *Political Metaphor Analysis: Discourse and Scenarios*. Bloomsbury Publishing.

Pavlovic, T. (2012). Cleanliness is Next to Godliness? Clean and Dirty Metaphors in English and Serbian. *ELALT*, 27–40.

Semino, E. (2008). *Metaphor in Discourse*. Cambridge University Press.

Sullivan, K. (2017). Conceptual Metaphor. In B. Dancygier (Ed.), *The Cambridge Handbook of Cognitive Linguistics* (pp. 385–406). Cambridge University Press.

7

The Politicians' Use of Mixed Metaphors in the U.S. Climate Discourse

The seventh chapter sheds light on mixed metaphor. Metaphors coming from two or more different source domains (not from one source domain as is the case with individual metaphors or extended metaphors) within the same clause. This leads to an overlap of entailments which in turn can direct the addressee to perceive realities that the speaker/writer wishes his/her addressee to perceive. This chapter discusses the linguistic restriction of mixed metaphor (within a single clause and in adjacent clauses). It also discusses the power of mixed metaphor. Based on my explanations and previous studies, I explained how mixed metaphor can have an effective impact on the addressee. An interesting finding regarding mixed metaphor is that all the mixed metaphors (regarding the data of governors and mayors), in this book, consist of the following pairs: Journey and War, Journey and Cleanliness, and Journey and Construction. Throughout the data, the journey metaphors are present in each of the mixed metaphors as the basic component of this mix. One example of a mixed metaphor (journey, war and cleanliness) from the data: "...Our states **will not stop fighting for a clean energy future**, regardless of who is in the White House..." the Governor of Washington's statement. This chapter thus explains the centrality of journey metaphor in mixed metaphor and how journey along with the other domains frame climate change. In this chapter, I drew a figure and framework representing the centrality of journey metaphors in mixed metaphor.

O. K. Al-Shboul, *The Politics of Climate Change Metaphors in the U.S. Discourse*,
https://doi.org/10.1007/978-3-031-19016-2_7

Mixed Metaphors

This chapter examines how and what metaphors are mixed and what source domains they are drawn from. This can show the policies that these politicians tried to make. Besides, using metaphors drawn from specific source domains reveals the nature of arguments made by the politicians, which centers on specific notions that shape the politicians' ideologies and shape the addressee's thinking of the social world (in this context, how they should think of climate change and/or Paris climate agreement and how they should react). Studying this mixing, I will highlight important themes in this chapter.

1. The linguistic restriction of mixed metaphor (Within a single clause; in neighboring clauses (with examples provided), where politicians draw on different source domains (e.g., journey and war) to make up a (mixed) metaphor within a single clause or adjacent clauses.
2. The power of mixed metaphor (stronger image; semantic) with example provided.

Politicians draw on various source domains when addressing an issue. Barnden (2016) explains that "mixing is usually taken to mean that the same target A is viewed both as B and as C more or less at the same time in a piece of discourse, with B and C being distinctly different source subject matters" (p. 76). This means that although metaphorical expressions can be adjacent to one another, they can be drawn from two or more quite different source domains to conceptualize the same target domain (e.g., climate change).

This chapter sheds light on metaphors coming from two or more different source domains (not from one source domain as is the case with individual metaphors in the two previous chapters) within one linguistic domain (the same single sentence). Mixed metaphors "use two or more different source domains to predicate something about the same target domain in a short stretch of discourse" (Forceville, 2016, p. 223). However, while some scholars (e.g., Charteris-Black, 2018) considered linguistic environment should only be within a single clause, others (e.g.,

Forceville, 2016) are more flexible, broadening the linguistic environment of mixed metaphors to include more neighboring clauses.

Kimmel (2010) states that metaphor clusters fulfil three kinds of functions: "First, they are attention-grabbing.... Second, clusters seem to occur where the action is. Finally, metaphor clusters connect and dynamize discourse" (p. 98). Many scholars discussed the notion of mixing (Kimmel, 2010; Lakoff & Johnson, 1980; Semino, 2008). The linguistic environment is important in determining this type of pattern, which is represented and identified by the clause borders. They set out grammatical boundaries for mixed metaphors: occur across adjacent clauses within the same sentence. "The mixing of metaphors creates a semantic inconsistency on the level of literal meaning" (Muller, 2016, p. 41), and one of the scenarios of this clash or inconsistency is that it can be subsumed under the same or different source domains (Semino, 2016). Identifying mixed metaphors has been beneficial in understanding the topic conceptualized by such mixed metaphors. In medical discourse, for example, Charteris-Black (2016) says that when the patient feels severe pain, he/she uses metaphors from different domains. That is, through experimental research he found that a patient draws on source domains that are widely different to describe how severe and complex the problem is. "The greater semantic divergence of metaphor source domains, the more intense the embodied experience of pain" (Charteris-Black, 2016, p. 155). Thus, an individual target domain can be conceptualized or conceived through different and unrelated source domains (Kövecses, 2010, 2016) for the purpose of helping the speaker/writer to better and more comprehensively describe or conceptualize the target domain to the addressee.

Naciscione (2017) argues that "it is believed that the issue of mixed metaphors has emerged over recent decades. However, it has existed in traditional stylistics for a long time" (p. 259). With regards to the predominant use of mixed metaphors '*mixaphor*' in political discourse, Kövecses (2016) says that "the target domains, or frames, ... have many different aspects to them, and these aspects normally require different source domains, or frames, for their conceptualization" (p. 6). According to him (Kövecses, 2016), many target domains have a wide scope. As a consequence, when a target domain is conceptualized, metaphorical expressions that are used for this conceptualization may not be homogenous. That is, they may not belong to the same source (semantic) domain but to different source domains. This means that the speaker or the writer may use two or more metaphors from different source domains in the course of producing discourse for the purpose of depicting what is addressed (the target domain). These metaphors are thus mixed (Charteris-Black, 2016; Kimmel, 2010). Moreover, Lakoff and Johnson (1980) argue that the speaker or the writer resorts to mixing since no single metaphor may focus on all aspects of a topic that he/she needs to highlight. They further maintain when it comes to mixed metaphors, entailments overlap as they come from different source domains. This overlap of such entailments can direct the addressee to perceive realities that the speaker/writer wishes his/her addressee to perceive. "Entailments make metaphor cognitively useful, because they provide new ways of thinking about the target domain" (Sullivan, 2017, p. 395).

This chapter throws light on this important type of metaphor, where two or more metaphors existing in a cluster or stretch come from different source domains (journey and war, for instance, as we will see in the analysis). Metaphors from different source domains contribute to making reality more prominent than when one metaphor (or a number of metaphors) from one source domain is used to conceptualize a topic. In this regard, Burgers et al. (2016) argue that "complex figurative frames could present an even stronger image than frames with a single figure. We therefore also propose that complex figurative frames boost frame salience (compared to frames based on a single figure)" (p. 423). That is, drawing on two or more source domains to describe one target domain gives a more comprehensive image than drawing only on one source domain to describe the same target domain since the image is shaped by using concepts drawn from more than one source domain. This means that the target domain is simultaneously described or conceptualized from more than one angle to give a clear picture of the target domain in ways that make the writer's argument more effective.

The following examples show two or more metaphors from different source domains (e.g., journey and war) mixed.

Example 1
We are also partnering with other states who are doing the same, and our states **will not stop fighting for a clean energy future**, regardless of who is in the White House. From the Governor of Washington's statement.

Here is a mix of metaphorical expressions used by the governor of Washington. The governor used three metaphorical expressions *will not stop*, *fighting for* and *a clean energy future* from three source domains, namely, Journey, War and Cleanliness. These expressions are all adjacent to one another, constituting one lexical unit whose function is to conceptualize how climate change should be tackled and also for what or why this problem should be tackled. That is, Washington and other states should keep responding to climate change by implementing effective measures to achieve their goal, which is to use clean (unpolluting) energy that the country and the world can depend on now and in the near future, regardless of what challenges they might confront.

Example 2
We will **continue building on our efforts** to address climate change. From the Governor of Connecticut's statement.

The target domain is conceived through two source domains. These domains are: journey and war. In the above example, the metaphorical main verb *continue* and the metaphorical gerund *building* noun are syntactically integrated in metaphor cluster complementing each other in a single (or closely connected) clause. The politician (the governor of Connecticut) uses two metaphorical expressions in a single clause to conceptualize the journey and construction aspects of (how to deal with) climate change since working constructively cannot be accomplished one time or over a short period of time. But it is constant and progressive if they really want to combat climate change. The lexical item *continue* functions as an adjective that describes how cooperation 'building' should be. That is, it needs to be progressive as long as climate change is happening. Using this mix of metaphor, the governor tries to present the nature of the problem, as being long-term and complicated. As a result, it should be dealt with by following steps and through cooperation.

Example 3

"**Today's move builds on action Congress** took to rebuff then-President Obama's regulatory rampage, which put American jobs at risk. When the previous administration signed America up for this unattainable mandate, we made it clear we would fight this unilateral action any way we could, and this day could not have happened soon enough. President Trump has once again put families and jobs ahead of left-wing ideology and should be commended for his action." The U.S. Senate Majority Leader Mitch McConnell.

Here the U.S. Senate Majority Leader Mitch McConnell, who is considered among politicians who supported the president when he decided to denounce the Paris agreement, uses mixed metaphors of journey and construction domains to construct an important meaning, which is this decision is a step in the right direction which contributes to building the country's economy through rejecting "Obama's regulatory rampage, which put American jobs at risk".

Then he uses another metaphor of war to tell the addressee that due to Obama's policies, they have been fighting for this unilateral action in their best ability. The ideology constructed in here is that Trump is rescuing the economy from devastating further. This entails that Paris agreement has destructive impacts on the economy. And the president now is trying to fix and rebuild what was damaged.

Example 4

Trump alienates allies, **walks away** from #ParisAgreement & **sabotages** health care. His presidency is all destruction & no accomplishments. Tim Kaine

Here Tim constructs this ideology that the president moves away from the agreement that he is supposed to keep. Tim indicates that the direction which is away from the Paris agreement is the wrong direction. Thus, the only right direction is moving into (rather than from) the Paris agreement. Here, the senator uses the phrasal verb *walk away* to connotate the negativity that the President made because of exiting the Paris climate agreement. He also uses another war metaphor "sabotages health care" linked with the conjunction 'and' with the previous verb (walks away) to construct another but related reality. This reality is that leaving the Paris agreement has disastrous effects.

Example 5
If the White House won't lead, California will step up. We will **continue our fight** against climate change. #ParisAgreement #ActOnClimate. Sen. Dianne Feinstein, D-Calif

The senator used two metaphorical expressions of war and journey to construct an image of resistance against climate change despite the decision made by the President. The senator using this mixed metaphor to show how strong people in California as they work so hard to deal with climate change. This metaphor reflects a steady and reluctant policy for dealing with climate change. These two expressions of metaphor drawn from two different domains contribute to creating a reality. That is, dealing with climate change will continue in spite of the U.S. withdrawal from the Paris agreement.

Example 6
Our track record-reducing carbon pollution while growing jobs and our economies- provides proof that we need not **sacrifice** opportunity for action. Indeed, we can **secure** that opportunity only by **continuing to lead**. From the Governors' letter to the President.

In this example, the two expressions *secure* and *continuing to lead* are metaphorical expressions representing two different source domains (War and Journey). These two metaphors located in one single sentence are used to highlight some aspects from these two different domains about climate change. The word *secure* preceded by the metaphorical expression *sacrifice* are terms that are frequently used in war situations. Here the governors recommend that the opportunity should be secured and not be lost at any way. Thus, in for the United States to secure the opportunity, then it should continue to lead. With these metaphorical expressions being used, aspects to do with War and Journey are highlighted and activated simultaneously in the addressee's mind to influence the way he/she can think of climate change and the way climate change should be addressed. In this way, the addressee can develop a sense of urgency (so that the opportunity not lost) to stay committed to the Paris deal or otherwise many damages might occur because carbon pollution and emission.

Example 7

Maintaining the U.S. commitment is essential to **protect** our residents, and indeed, all Americans from the potentially **catastrophic impacts** of a changing climate. From the Governors' letter to the President.

The first two metaphorical expressions *maintaining* and *protect* belong to two different source domains: Journey and War respectively. Another metaphorical expression *catastrophic impacts*, drawn from the source domain of war, follows these two expressions. It instantiates the consequences of this problem. However, the first two metaphorical expressions are almost adjacent, occurring in close proximity to one another within the (single) sentence. The sentence starts with the gerund verb *maintaining* (the metaphorical expression) where the governor indicates the reason for the U.S. commitment. Also, the metaphorical expression *protect* is used to indicate why the U.S. commitment is important. Both the metaphorical expressions *maintaining* and *protect* occur within a main syntactic unit (sentence) in a cause-effect relationship. This syntactic relationship contributes to constructing the message to the addressee. The mixing here is functional. That is, it can trigger inferences that the writers want their audience to make and perceive. By way of explanation, journey aspects overlap with war aspects forming a message that staying in the Paris agreement is a necessary step to save the Americans from all sorts of dangers associated with climate change.

Example 8

In each of our states, the **path forward** is **clear**. Our citizens demand the low-cost, clean-air benefits. From the Governors' letter to the President.

The writers interweave metaphors from two different source domains (Journey and Cleanliness) in a way that strengthens their stance on the issue. The two expressions altogether form a mixed metaphor (since they are adjacent, yet they belong to different source domains). In this example, aspects from the Journey and Cleanliness domains are highlighted. They merge and blend to draw an image in the addressee's mind that the states each have well-planned strategies by setting up projects for the purpose of using clean energy. The lexical item *the path forward and clear* assert that states have already started using nonpolluting resources.

Example 9
America's cities will **continue to lead the way** in **moving forward** in **protecting** our residents from the **disastrous** effects of climate change, and creating a thriving 21st century economy. From the Mayors' letter to the President.

Here in this example the writers formulate their vision towards dealing with climate change as cities using four metaphorical expressions *continue to lead the way*, *moving forward*, *protecting*, and *disastrous* which come from two different sources: journey and war. Aspects of the two different domains (courage, urgency, confronting and progress) are highlighted to conceptualize the target domain in the addressee' mind so that the mayors' message is constructed in ways that they want the addressee to perceive. As can be seen here, the mixing of these metaphors reveals that the U.S. cities should lead by implementing measures to tackle the problem. This is the only path which they need to go through or otherwise the residents would be at the risk of the consequences of climate change (because protection from those consequences in such a case is not guaranteed).

Example 10
Our leading U.S. companies recognize the need to address **business risks** and opportunities **through** the Paris Agreement, and are wisely investing in low-carbon fuels and technologies to stay on the cutting edge of the world economy. From the Governors' letter to President.

In this example two metaphorical expressions, drawn from the domains of War and journey, were used: *business risks* and *through*. They both occur in close proximity, which can be identified as transitive verb followed by object then by a preposition. The journey metaphor (through) is used to conceptualize the way through which climate change should be addressed; which should be via the Paris agreement. With this mixing (War and Journey), the governors in their letter see this agreement as the path through which the U.S. companies recognize the need to confront and deal with the consequences (business risks). The preposition *through* carries the meaning of indicating direction that the U.S. companies need to follow for the purpose of dealing with climate change, and correspondingly the Paris agreement is conceived as a necessary step that the U.S. should follow (when discussing climate change in terms of risks and opportunities).

Example 11

And if the President wants to break the promises made to our allies enshrined in the historic Paris Agreement, we'll **build and strengthen** relationships around the world to **protect the planet from devastating climate risks**. From the mayors' statement.

Here, in the example, the Construction metaphor and War metaphor are mixed. The mayors used metaphorical expressions from these two domains: *build, strengthen, protect, devastating climate risks*. These metaphors of Construction and War domains are stated in the cause and effect order in a single sentence. Dealing with climate change requires implementing measures starting with strengthening the local and world stance by building more relationships between the country and many other countries around the world (and the Paris agreement is one great way to build such relationships). The means and goal are conceptualized by the Construction and War metaphors, realized by the lexical items: *build* and *strengthen* for the Construction metaphors and *protect, devastating* and *risks* for the War metaphors. This mixing of metaphors made the mayors' message or meaning more powerful in the sense that it directs the addressee's thinking of how the world should be protected (which is by keeping strong relationships with the world). That is, if we need to make the world a safe place, then we need to cooperate so that our relationships become stronger, and consequently we can effectively tackle this problem (climate change).

Example 12
If the U.S. does not maintain world climate leadership through national policies to reduce greenhouse gas emissions and **transition to clean** energy, China and India will. From the Governors' letter to the President.

In this example there are two metaphorical expressions *transition* and *to clean energy* from the Journey and Cleanliness domains. Both of which occur in very close proximity (being adjacent to one another in the first conditional clause). Since they are drawn from two different domains (Journey and Cleanliness), they are regarded mixed metaphors. In this example, the governors emphasize the necessity that the U.S. is as always should be the leader and this role in turn requires them to move forward, not backward, through using unpolluting resources. The governors aim at keeping the US's stance on the federal level in the right track. That is, the U.S. must show its leadership by staying in the Paris Agreement and the withdrawal undermines its position as a leader. Aspects of source domains of Journey and Cleanliness are activated to generate new conceptualizations about the target domain. Here the governors try to affect the decision makers (the President) that the U.S. should always be a leader in this field or otherwise this position will be taken by others. A similar mix existing in the data is "a clean energy transition" where aspects of journey and cleanliness are mixed. That is, the right path is determined by the transition from polluting into non-polluting energy.

Lakoff and Johnson (1980) and Rigney (2001) argue that sometime a single metaphor might not sufficiently do the job (conceptualization and facilitation). In other words, two or more metaphors are needed to describe one target domain such as climate change especially that the issue is rather abstract and complex and above all no social consensus from the politicians is on climate change. Cameron (2016) argues that "the use of multiple metaphors in human understanding is necessary and inevitable; that our multiple metaphors sometimes seem to clash is mostly an illusion perceived from outside of discourse" (p. 29). Accordingly, the metaphor mixing is functional for two reasons: (1) It provides aspects necessary to frame the problem that needs to be addressed. (2) It simplifies (facilitates) presuppositions. In this regard, Charteris-Black (2011, 2014, 2018) maintains that when metaphors are made of two or more source domains, they tend to be more effective than those that are made of only one source domain because the speaker/writer can shape his/her argument in a more comprehensive way, and thereby he/she can convince the addressee.

But it is also worth investigating which source domains from which the mixed metaphors are made. In the data collected and analyzed, mixed metaphors consist of two or three metaphorical expressions from two different source domains existing in a single sentence. I found that almost all the mixed metaphors, in this book, consist of Journey and War, Journey and Cleanliness, and Journey and Construction. Throughout the data, it is noticed that the journey metaphors are present in almost all the mixed metaphors in this book, as the basic component of this mix.

Figure 7.1 shows journey as a central domain around which the remaining domains revolve.

This leads us to an important question: Why is the Journey metaphor a basic component in these mixed metaphors? This finding demonstrates that climate change is basically conceptualized as Journey. But how can this be explained? The politicians tend to conceptualize the nature of the problem and what solutions should be offered accordingly in terms of journey. Among the most prominent characteristic of climate change is

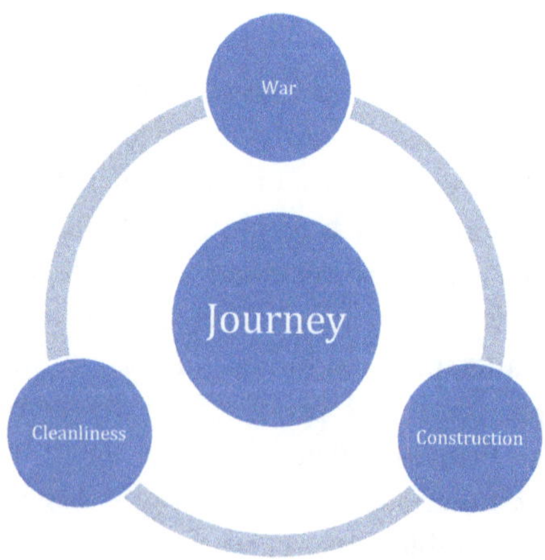

Fig. 7.1 The centrality of journey amongst the other three types of metaphors

that it is a long-term phenomenon where the focus is not on destination itself but on movement in the right direction towards the destination according to many scholars (e.g., Atanasova & Koteyko, 2017; Semino, 2008). Thus, the focus is on how to deal with this long-term problem; and correspondingly what appropriate measures and procedures needed to tackle this problem. This explains why the politicians here in this book heavily used metaphorical expressions from the Journey domain and why Journey metaphors are the basic component in all the metaphor clusters.

The mixed metaphors of Journey and War make up most of these mixed metaphors in the data. This leads us to investigate why most mixed metaphors come from these two domains in particular although Journey and War appear to be contradictory since war emphasizes immediate initiation, and journey, conversely, emphasizes following steps (though the steps might be swift). One way to explain this is that because the problem is threatening and serious, the governors and mayors stress that the governments should start to address climate change promptly to protect the environment from damage and to avoid the consequences of this problem (as conveyed by the War metaphors). Then, once fighting climate change starts, the focus then should be on how climate change needs to be dealt with (through initiating and managing many long-term projects).

In such mixed metaphors, the most prominent notions of the source domains are highlighted. That is, while War prompts action, journey emphasizes following steps. Besides, there are other explanations of why many metaphors are drawn from these two source domains (Journey and War): War and Journey seem to be very familiar domains, which encourage the politicians to draw on them in a mixed style. Also, the nature of the problem (climate change) led these politicians to mix War and Journey metaphors in a way that make their message more effective as the problem is long-term and dangerous. Therefore, War metaphors are used to alert the addressee about the seriousness of the problem (Atanasova & Koteyko, 2017) and being at War needs patience and steps to be followed and no metaphor can better conceptualize the steps-based solutions than Journey metaphors. Consequently, Journey metaphors are mixed with war. It can thus be argued that although War and Journey metaphors seem quite different domains, they are complementary to one another.

It is worth mentioning that it is not necessary that all aspects associated with one source domain to be highlighted when it is used to describe a topic. In other words, the nature of the topic that is described using metaphors from the source domain determines to a large extent which aspects or notions that can be activated and which are not. For example, while journey comprises concepts such as progress, movement and destination, Journey also involves concepts of entertainment, joy and hanging out, but these concepts are not highlighted when the speaker draws on the domain of Journey to describe such a threatening-life problem as climate change. These concepts (entertainment, joy and hanging out) can be highlighted when journey is used to describe another topic that is different from climate change. For example, running a business or pursuing higher education is journey: running a business or pursuing higher education is serious but not life-threatening. Thus, concepts such as entertainment and joy can be activated when a speaker/ writer describes a running a business or pursuing higher education with journey metaphors. Simply put, climate change itself is a danger, but business or education is not. Accordingly, when climate change is described in terms of journey, what is crossed with the topic (climate change) and with the domain of War is not those aspects (entertainment, joy and hanging out) at all. That is, the aspects of journey that are activated along with climate change and the domain of War are *steps*, *progress*, *the right direction* and *destination*. These are the ones that contribute to forming the speaker's meaning, which the addressee perceives as a reality, when he/she talks about such a topic as climate change.

The second group of mixed metaphors is Journey and Construction. This can be explained by the fact that building a house, for instance, is something difficult and requires many workers so they can construct a well-structured base upon which they can build a strong house. In this mixing, construction aspects convey the meaning that efforts (implementing measures and establishing projects) should be strong so they can be effective and these efforts should be all exercised in dealing with climate change, which represents a movement forward in the right direction. Thus, construction emphasizes cooperation to strengthen efforts through employing measures to address the problem, starting at making a base and then going through many steps. As a result, Journey metaphors are employed to emphasize following steps to achieve progress and to continue building efforts (starting from building the basis to having a well-structured house). This kind of source-to-target mappings can invite the addressee to some presuppositions: many efforts from many policy makers and also people are required, and these efforts cannot be exercised within a short period of time.

For Journey and Cleanliness mix of metaphors, this mix can be explained as cleanliness represents one of the important goals that the governments and countries hope to reach through following steps. These steps are conceptualized by the Journey metaphors. The governors and mayors call for implementing more and more measures and for initiating projects to deal with this problem (climate change) so that the country can transform to using this new type of energy (clean energy).

Table 7.1 gives what each of these mixed metaphors entails, based on the conceptual metaphors that each consists of.

Burgers et al. (2016) argue that a conceptual metaphor "often implies a story and/or event sequence enabling metaphors to function as reasoning devices" (p. 4). This story or sequence of events depicted in conceptual metaphors can be captured as a problem and solution. This is consistent with what Landau and Kefeer (2014) argue:

> Metaphor use also helps people to think through abstract problems and their potential solutions. When a problem is vague, abstract, complicated, or poorly defined, people may not be sure what actions are available, the various possible outcomes of those actions, or how to weigh the costs and benefits of different options. Metaphor use allows people to use knowledge of a familiar scenario to reason about possible actions to take in the target situation, what goals to take into account, and how to forecast the potential effects of their actions. Because each metaphoric mapping highlights and downplays select features of a target, using different metaphors for the same target problem can orient people toward different solutions. (p. 465)

Table 7.1 The main entailments of each of the mixed metaphors

Journey and War	Journey and Construction	Journey and Cleanliness
To protect people and the environment from the consequences of climate change, the policy makers should promptly take action to begin fighting climate change. Once fighting starts, it takes long time to reach the destination since tackling the problem requires complex and long-term measures.	Dealing with climate change needs building as many efforts on the part of policy makers and people as possible, and these efforts and measures should be in the right direction to reach the destination.	Procedures and measures need to be implemented over a period of time so that the country can transform into and use clean technology/ power in the country as an alternative choice to the conventional type.

Table 7.2 summarizes how the four conceptual metaphors (Journey, War, Cleanliness and Construction) frame climate change in the data analyzed, where these four conceptual metaphors are realized. In this framing, the focus is on climate change as being a problem and at the same time on what the solutions are proposed to this problem accordingly. Based on the four conceptual metaphors (Journey, War, Cleanliness and Construction), which are realized from the metaphorical expressions employed by the governors and mayors, the characteristics of the problem (climate change) and solutions to this problem are identified, as the Table 7.2 indicates.

Table 7.2 The entailments of the conceptual metaphors in terms of cause and effect

	Types of Metaphors	Problem	Solution
1	Journey	long-term	steps
2	War	threatening	urgent/strict
3	Cleanliness	Man-made	new or creative measures
4	Construction	difficult	unity and cooperation

Through the metaphors used, at the macro level, the governors and mayors framed climate change in ways that make it appear as a problem with specific characteristics, and therefore specific solutions are required. By way of explanation, with regards to describing and framing this problem using the four types of metaphors, it can be argued that this problem and solutions offered can be formulated in cause-and-effect formulas. That is, since the problem is framed as long-term, the solutions should focus on following steps; since the problem is framed as threatening, effective measures should be promptly implemented; since the problem is framed as Man-made due to emissions, the solutions should be relying on alternative (clean) resources; and since the problem is framed as complicated and difficult to deal with, the solutions should focus on unifying and strengthening the efforts so that they become more effective to tackle the problem.

An important aspect related to mixed metaphors is formality. That is, drawing on figurative devices is among aspects that can reflect formality in language produced. For example, metaphors, hyperbole and simile can increase the level of formality in writing. When a piece of writing involves a higher number of such language features, it is characterized with a higher level of formality than a piece that includes no or only few of these features.

Mixed metaphors are more likely to be used in written forms rather than in spoken forms since formal writing gives writers longer time to draw on language features so that it can involve more complex features (e.g., mixed metaphors) since producing metaphors consisting of two or more different source domains is more difficult than using metaphors from only one source domain in a single clause. At the cognitive level, this difficulty can be attributed to the fact that the speaker needs to find metaphorical ties from these source domains to match with the topic (the target domain) addressed in ways that the addressee can be able to inferentially process this metaphorical meaning to solve semantic inconsistency. This gives explanations of why mixed metaphors are used in the letters (formal form of writing) much more than in the statements (informal form of writing). Based on that, it can be argued that mixed metaphors (two-or-more-source domains metaphors) can be seen as one aspect that reflects complexity, and hence formality more than one-source domain metaphors.

As indicated in Table 7.3, the politicians used mixed metaphors in their letters more than in their statements. This can be attributed to the fact that language in the letters is more complex than statements since these letters were formally written and given to the president. However, the statements were written and published online. Language, therefore, tends to be less complex. Accordingly, it can be argued that mixed metaphor is one way of reflecting complexity and formality in writing. In conclusion, a text with many metaphors reflects a higher level of complexity than a text with few metaphors. Also, a text involving many mixed metaphors can be more complex and formal than a text involving only individual metaphors or fewer mixed metaphors. Therefore, mixing can be one aspect which reflects formality in language.

Mayors altogether issued a statement, while (the seventeen) governors each issued an individual statement. One way to explain this is that the governor occupies a far higher position of authority at the executive level so that he/she can speak on behalf of (the larger population) the state he/she governs. Mayor's voices, however, would be stronger when they aggregated so that they can altogether construct a stronger voice, due to the fact that unlike the governor, the mayor governs a city or district. This

Table 7.3 The number and percentage of mixed metaphors out of all the metaphors used in the statements and letters

Genre	The number of all metaphors in the data	The number of mixed metaphors	Percentage of mixed metaphors out of the total metaphors
Statements	141	38	26.9%
Letters	57	32	56.1%

means a small proportion of population that the mayor governs in comparison with that of the governor. For the letters, 12 governors wrote a letter to the President. Also, 57 mayors gave a letter to the President responding to the decision. In these letters they expressed their opposition to decision and they called for addressing climate change. Semino (2005) maintains that metaphors that are used by a group can express a perspective that might influence the addressee more than those ones that are used by an individual.

Another aspect that can be discussed through mixed metaphor is urgency. Since the politicians (the senators, governors and mayors) use metaphors to urge action on climate change as they oppose the U.S. withdrawal from the Paris climate agreement. This section investigates urgency and which metaphors stress stronger urgency than the remaining. It investigates the intensity of urgency of metaphors to respond to climate change, which is an important way to make a distinction between the four conceptual metaphors. In this section, I developed a model which can be used to measure urgency of these metaphors, based on a scale, so that I can examine which regions and which politicians stress a stronger urgency than the others.

I evaluated the conceptual metaphors based on the following: war stresses a stronger urgency than the remaining conceptual metaphors. That is, war demonstrates how serious and threatening a situation is. This domain (war), therefore, foregrounds the idea that climate change should be dealt with promptly (so implying stronger urgency), as argued by Atanasova & Koteyko, 2017; Flusberg et al., 2017, 2018; Nerlich, 2009; Yang, 2017. Journey involves urgency (and it can be fast and urgent). But here the focus is on what the initial perception of the addressee is when he/she hears or reads war or journey metaphor. Journey focuses on following steps, meaning step 1, then 2, 3, 4 etc,. These steps can be very fast. But while war involves steps or stages, our initial perception not on the necessity that steps need to be followed in sequence but on the necessity of starting step number one, which is in the context of this book triggering an immediate and serious action against climate change. We are thus talking about our initial perception (about what it comes first to our mind when such metaphors from such domains as war or journey are used). Flusberg et al. (2018) explain why war stresses stronger urgency than other metaphors. They state that "the vivid emotional valence associated with war can quickly activate a sense of urgency and anxiety, which may motivate further action under some circumstances (at least in the short term)" (p. 11). They (Flusberg et al., 2018) also maintain that War metaphor causes fear and anxiety and communicate threat to people from the target domain. "This fear can motivate people to pay attention, change their beliefs, and take action about important social issues" (Flusberg et al., 2018, p. 6), and thereby evokes a sense of urgency. Politicians and journalists increasingly employ metaphors of War to stress the urgency of dealing with climate change so that urgent actions can be taken to tackle this problem (Cohen, 2011). Journey at the other side of the scale emphasizes taking steps to tackle this problem (so less urgency) as argued by many researchers (e.g., Atanasova & Koteyko, 2017; Fløttum, 2017; Semino, 2008). That is, in Journey metaphors, the focus is more on movement to destination rather than destination itself (Atanasova & Koteyko, 2017). Cleanliness and construction are in between the two domains. However, cleanliness implies a stronger urgency than the construction domain since cleanliness emphasizes offering a clear and direct solution. Simply put, the addressee is already introduced to choices offered by the speaker (which is using clean energy as an alternative to the conventional energy). The domain of construction, on the other hand, focuses more on the process of unifying efforts so that they policy makers cooperate to offer solutions to deal with the problem, and thereby these efforts can be effective. The concept of construction "highlights the importance of ongoing support—a foundation is just a start, and further construction and maintenance are needed to create a solid, stable structure that endures" (frameworksinstitute.org/pubs/mm/talkinghumanservices/page7.html). Accordingly, I ordered the conceptual metaphors in the following scale (Fig. 7.2) starting from which metaphor stresses highest to the lowest level of urgency: War [40], Cleanliness [30], Construction [20] and Journey [10], from highest to lowest urgency.

According to this scale (Fig. 7.2), War metaphor has the highest level of urgency, then followed by Cleanliness and then by Construction. For Journey metaphor, it has the lowest level of urgency.

I used this scale (Fig. 7.2) to measure metaphors in terms of urgency for two reasons: first, metaphors in this book belong to four conceptual

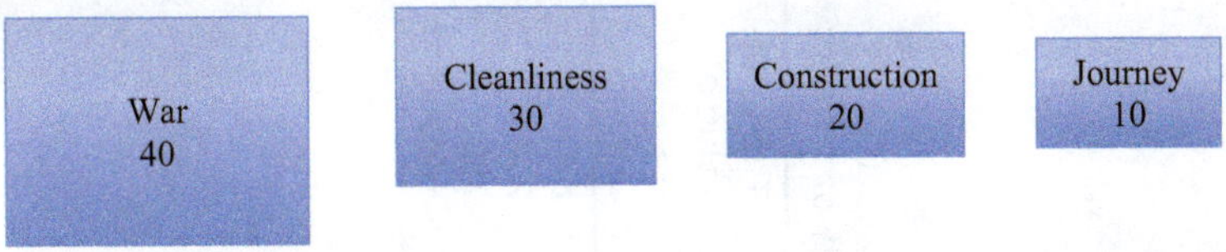

Fig. 7.2 The scale of the urgency of metaphors

metaphors and this scale is one way to distinguish between these metaphors, based on which type of these metaphor stresses relatively stronger urgency than the others. Secondly, this scale is one way to show which governors stress a stronger urgency and which regions in the U.S. call for prompting action on climate change more than the others when the President made the decision. For example, when a governor uses War metaphor, he/she emphasizes a stronger urgency than other types of metaphors used.

Table 7.4 gives information about the frequency of each of the metaphor types in each region. The Table 7.4 also shows the value of urgency. This value is calculated based on what the metaphor is and the frequency of each type of the metaphors (these values are shown in Fig. 7.3).

Table 7.4 and Fig. 7.3 indicate the number of metaphors that the governors in each region used and also, thereby, the total value of urgency. We can thus determine which region stressed a stronger urgency than the other. In this way I can determine which regions are more willing to act on climate change and offer some explanations of these results. Since the number of statements of the Northeast and West is equal and Midwest and South almost equal, I compared Northeast to West (six governors to six governors) and also Midwest to South (two governors to three governors). It can be clearly seen that the West and Northeast stressed urgency, but governors of West expressed stronger urgency than Northeast, and South stressed stronger urgency than Midwest.

Additionally, I counted the number of governors in each region in the country who opposed the decision to find which of these regions reacted more than the others.

Table 7.5 shows the number of states in each region that opposed the decision. In the Table 7.5, it can be clearly seen that the two thirds of governors in the Northeast responded to the decision. Partial statistics is

Table 7.4 The number of metaphorical expressions by the governors of each of the regions and the intensity of urgency

The official regions	Journe y metaphors	The value of urgecy	War metaphors	The value of urgency	cleanl-iness metaphors	The value of urgency	Construction metaphors	The value of urgency	The total value of the urgency
Northeast	13	130	10	400	5	150	6	120	800
Midwest	2	20	4	160	1	30	1	1	210
South	13	130	13	520	1	30	1	20	700
West	18	180	24	960	14	420	2	40	1600

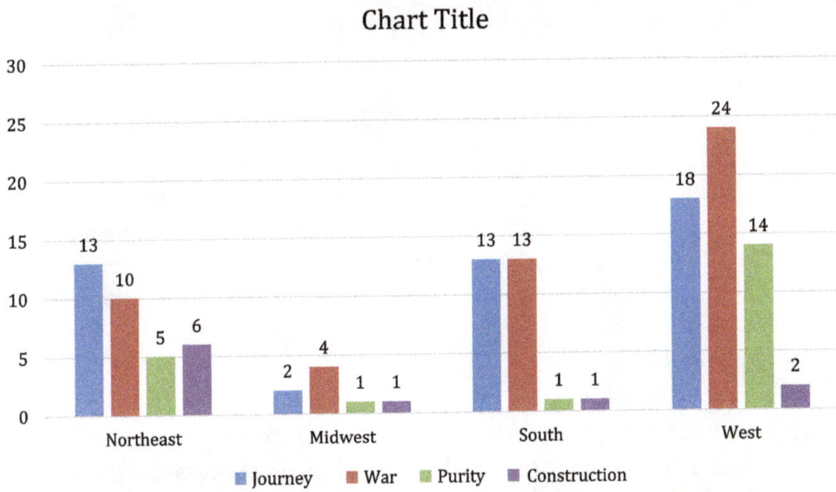

Fig. 7.3 The distribution of metaphors by the governors in the four regions

Table 7.5 The number of the governors opposed the decision in each region

The official regions	The number of states in the region	The number of the states opposed withdrawal	The percentage of the states who opposed withdrawal
Northeast	9	6	66.66
Midwest	12	2	16.66
South	16	3	18.75
West	13	6	46.15

conducted to determine the percentage of the governors out of each region for the purpose of identifying which region has reacted more than the others. Based on the proportion of the states which opposed the decision to the proportion of states which each region includes, I found that more governors in the Northeast reacted against the decision than in any other region in the country: That is, two-thirds of the Northeast governors (six governors out of nine) opposed the decision. This leads us to a further investigation of why the Northeast was the most reacting region to the President's decision of withdrawing the U.S. from the Paris agreement. This can be attributed to the fact that the Northeast states are more affected by climate change than the other regions.

There are many more risks that the population in this region (Northeast) encounters:

> Changing climate threatens the health and well-being of people in the Northeast through more extreme weather, warmer temperatures, degradation of air and water quality, and sea level rise. These environmental changes are expected to lead to health-related impacts and costs, including additional deaths, emergency room visits and hospitalizations, and a lower quality of life. (From fourth national climate assessment report, 2018, p. 697, https://nca2018.worldchange.gov/downloads/NCA4_2018_FullReport.pdf)

This explains why more governors responded to the threats of climate change and reacted against the decision than governors in the other regions in the country. "The Northeast region has traditionally been a leader in greenhouse gas mitigation action, serving as a potential model for other states" (From fourth national climate assessment report, 2018, p. 695).

The West comes in the second place where about half of its governors reacted to the decision. It has the second highest percentage (46.15%) after the Northeast. "Parts of the Southwest reach the hottest temperatures on Earth, with the world record high of 134°F (57°C) recorded in Death Valley National Park, California and daily maximum temperatures across much of the region regularly exceeding 98°F (35°C) during summer" (nca.worldchange.gov, 2018, p. 1108). This explains why most governors who opposed the decision are from the Northeast and West. "Communities on the front lines of climate change experience the first, and often the worst, effects" (From fourth national climate assessment report, 2018, p. 1038).

Metaphor as a Trigger for Social Change

In the light of the data analyzed, I discuss how the politicians (the President, senators, governors, and mayors) accessing to public resources of power use language (metaphorical choices) to trigger action on climate change and make social change, starting from how power is used to how social change is triggered. I drew on the main concepts which make up CDA analysis to explain how social change can occur, as the politicians (the President, governors and mayors) use linguistic features (metaphors) in this book to trigger social change.

Power is enacted in texts and exercised by politicians (the President, senators, governors and mayors) with varying powerful resources. "Power is discursively exerted not only by grammatical forms, but also by a person's control of the social occasion by means of the genre of a text, or by the regulation of access to certain public spheres" (Wodak, 2014, p. 4). Accordingly, with regards to the social actors in this book, the President delivered his speech from the White House when he officially announced that he would withdraw the U.S. from the Paris climate agreement. The senators, governors and mayors, on the other hand, issued their statements through the formal websites of the states and cities. Thus, the White House and formal (states & cities) websites are considered certain public resources that very few people (the President, governors and mayors in this context) can access to. These are important resources of power that the social actors have in this book. But the public resource that the President can access to (which is The White House) is more powerful and effective than the public resource that the governors and mayors can (which is the formal websites and other formal domains). Texts produced via these resources, however, are effective.

For the texts (the statements or letters) analyzed in this book, they are interrelated, which is one of the text properties. The analysis of metaphorical patterns reveals that intertextuality constitutes an important aspect of those texts (statements and letters) that can be realized in the data. For example, although each statement was separately issued by a governor, there are similarities with regard to the linguistic surface of these statements. This is evidenced through using similar metaphorical expressions which I found in these texts produced by senators, governors and mayors (e.g., fight climate change, clean energy and build efforts). In this regard, the existence of similar linguistic traces can be explained from the discourse historical approach (from the historical perspective). That is, texts almost are simultaneously produced to respond to the same occasion, which is the problem of the U.S. federal government's withdrawal from the Paris agreement in 2017. This emerging situation (the president's announcement to withdraw the U.S. from the Paris climate agreement) prompted politicians, including the governors and mayors to respond and oppose this decision. This can offer an important explanation of the intertextuality of the texts analyzed, which is an important aspect of texts according to several scholars (e.g., Fairclough, 2014). As a consequence, these texts are of dialogic nature, (which means that they are constructed in accordance with other texts) rather than of monologic nature.

This demonstrates that there is a social struggle between politicians: "Many environmental issues involve power struggles between opposing groups, and these struggles frequently take place in, and over, language" (Haig, 2001, p. 209). For example, the President versus the governors and mayors, with each trying to persuade the addressee (public) that what the President or the governors and mayors say is legitimate, and thereby making their opinions presented to the addresses as realities which are lexicalized by many metaphorical expressions (for the President: withdrawing from the Paris agreement is something necessary since it is not fair; for the governors and mayors: climate change needs to be urgently addressed since it is dangerous, and Paris agreement is a good opportunity to seriously deal with this problem). This means that the politicians need to persuade the addressee that what they say is right. Persuasion is thus the main goal of why texts, involving linguistics means, are produced. Persuasion is "a multi-layered discourse function that is the outcome of a complex interaction between intention, linguistic choice, and context" (Charteris-Black, 2005, p. 30).

It is worth asking why these politicians aim to present their opinions and arguments as realities. The president, senators, governors and mayors aim to trigger social change through presenting their realities or myths via linguistic means such as metaphor. "Metaphor…is a central rhetorical figure in creating a myth appropriate for the political purpose of motivating social action" (Charteris-Black, 2018, p. 264). But what is the social change needed from the governors' and mayors' perspective and they aim to trigger in the social world? In the context of this book, social change that is aimed to be made by the state and local governments using metaphor is to trigger the federal government's and people's *willingness of action*. I borrowed this term (willingness of action) from Fløttum (2017) to refer or define what social change that the politicians try to make. "The success of the Paris agreement is dependent on every country's implementation of the obligations stated in the document, and such implementation rests to a large extent on the citizens' willingness to act" (Fløttum, 2017, p. 113). Thus, *willingness of action*, conceptualized by metaphors from Journey, War, Cleanliness and Construction is to implement serious measures to tackle climate change starting at keeping joining the Paris climate agreement, as being a necessary step to deal with climate change. Another related social change that the politicians are trying to make is eliciting people to call for more measures and make more pressure on the federal government to act on climate change. They, therefore, aim at raising people's awareness of the current and future risks of climate change and all dangers associated with this environmental problem, as stated in the statements issued publicly by the governors and mayors. Such awareness aims to lead the public to push policy makers to address this environmental problem. On the other hand, the President's social change that he makes is to withdraw the U.S. from the Paris climate agreement (Table 7.6).

Each side holds an argument, and the argument, in turn, can be foregrounded or backgrounded, with the result that the audience may or may not accept that argument.

Table 7.6 The main groups and social change

The participants	Focus	Social change
The President &many politicians (senators)	The Paris agreement is not fair.	Withdrawing from the Paris agreement.
Governors, Senators & Mayors	Climate change needs to be urgently addressed; it is a real threat.	Maintaining in the Paris agreement as one important way to deal with climate change and calling for more actions on climate change.

References

Atanasova, D., & Koteyko, N. (2017). Metaphors in Online Editorials and Op-ed About Climate Change, 2006–2013: A Study of Germany, the United Kingdom, and the United States. In K. Fløttum (Ed.), *The Role of Language in the Climate Change* (pp. 71–89). Routledge.

Barnden, J. (2016). Mixed Metaphors: Its Depth, Its Breadth, and a Presence-based Approach. In R. W. Gibbs (Ed.), *Mixing Metaphor* (pp. 75–111). John Benjamins Publishing Company.

Burgers, C., Konijn, E. A., Steen, G., & J. (2016). Figurative Framing: Shaping Public Discourse Through Metaphor, Hyperbole, and Irony. *Communication Theory, 26*(4), 410–430.

Cameron, L. (2016). Mixed Metaphors from a Discourse Dynamics Perspective. In R. W. Gibbs (Ed.), *Mixing Metaphor* (pp. 17–30). John Benjamin Publishing Company.

Charteris-Black, J. (2005). *Politicians and rhetoric. The persuasive power of metaphor*. UK: Palgrave Macmillan.

Charteris-Black, J. (2011). *Politicians and rhetoric. The persuasive power of metaphor*. UK: Palgrave Macmillan.

Charteris-Black, J. (2014). *Analyzing political speeches: Rhetoric, discourse, metaphor*. First edition. UK: Palgrave Macmillan.

Charteris-Black, J. (2016). The Dull Roar and the Burning Barbed Wire Pantyhose: Complex Metaphor in Accounts of Chronic Pain. In R. W. Gibbs (Ed.), *Mixing Metaphor* (pp. 155–176). John Benjamin Publishing Company.

Charteris-Black, J. (2018). *Analyzing Political Speeches: Rhetoric, Discourse and Metaphor* (2nd ed.). UK: Palgrave Macmillan.

Fairclough, N. (2014). *Language and power*. 3rd edition. Routledge.

Fløttum , K. (2017). Willingness of action. In K. Flottum (Eds.), *The Role of Language in the Climate Change Debate* (pp. 113–129). New York: Routledge.

Flusberg, S., Matlock, T., & Thibodeau, P. (2017). Metaphors for the War (or Race) Against Climate Change. *Environmental Communication, 6*, 769–783.

Flusberg, S., Matlock, T., & Thibodeau, P. (2018). War Metaphors in Public Discourse. *Metaphor and Symbol, 33*(1), 1–18.

Forceville, C. (2016). Mixing in Pictorial and Multimodal Metaphors. In R. W. Gibbs Jr. (Ed.), *Mixing Metaphor* (pp. 223–239). John Benjamins PublishingCompany.

Haig, E. (2001). A Study of the Application of Critical Discourse Analysis to Ecolinguistics and the Teaching of Eco-literacy. *Critical Discourse Analysis for Ecolinguistics, 2*, 205–226.

Kimmel, M. (2010). Why We Mix Metaphors (and Mix them Well): Discourse Coherence, Conceptual Metaphor, and Beyond. *Journal of Pragmatics, 42*, 97–115.

Kövecses, Z. (2010). *Metaphor: A Practical Introduction* (2nd ed.). Oxford University Press.

Kövecses, Z. (2016). A View of Mixed Metaphor Within a Conceptual Metaphor Theory Framework. In R. W. Gibbs (Ed.), *Mixing Metaphor* (pp. 3–15). John Benjamins Publishing Company.

Lakoff, G., & Johnson, M. (1980). *Metaphors We Live By*. The University of Chicago Press.

Landau, M. J., & Keefer, L. A. (2014). This is like that: Metaphors in public discourse shape attitudes. *Social and Personality Psychology Compass, 8*(8) 463–473.

Muller, C. (2016). Why Mixed Metaphors Make Sense. In R. W. Gibbs (Ed.), *Mixing Metaphor* (pp. 31–56). John Benjamins Publishing Company.

Naciscione, A. (2017). Extended Metaphor in the Web of Discourse. In R. W. Gibbs (Ed.), *Mixing Metaphors* (pp. 241–266). John Benjamins Publishing Company.

Nerlich, B. (2009). The Post-antibiotic Apocalypse and the War on Superbugs: Catastrophe Discourse in Microbiology, Its Rhetorical Form and Political Function. *Public Understanding of Science, 18*, 574–590.

Rigney, D. (2001). *The Metaphorical Society: An Invitation to Social Theory*. Rowman & Littlefield Publishers.

Semino, E. (2005). The Metaphorical Construction of Complex Domains: The Case of Speech Activity in English. *Metaphor and Symbol, 1*, 35–70.

Semino, E. (2008). *Metaphor in Discourse*. Cambridge University Press.

Semino, E. (2016). A Corpus-based Study of Mixed Metaphor as a Metalinguistic Comment. In R. W. Gibbs Jr. (Ed.), *Mixing Metaphor* (pp. 203–220). John Benjamins Publishing Company.

Sullivan, K. (2017). Conceptual Metaphor. In B. Dancygier (Ed.), *The Cambridge Handbook of Cognitive Linguistics* (pp. 385–406). Cambridge University Press.

Wodak, R. (2014). Politics as Usual: Investigating Political Discourse in Action. In J. P. Gee & M. Handford (Eds.), *The Routledge Handbook of Discourse Analysis* (pp. 525–540). Routledge.

Yang, M. M. (2017). *American Political Discourse on China*. Routledge Studies in Rhetoric and Communication.

8

A Corpus-Based Analysis of Climate Discourse in the Web News

In the eighth chapter, I used the NOW (News On the Web) program to conduct a corpus analysis in specified periods. This program (NOW) has many search features. I used Collocate, Frequency and Compare tools available in this program (News On the Web) to investigate the main questions in this chapter. The researcher can refine the period of time he/she searches in (through using the search by date feature) and also which country he/she needs to search the data: the United States, Canada, France, Australia, etc. For selecting the U.S. corpus, I refined the search through sections (where I selected the U.S. corpus). The NOW corpus comprises web-based newspapers and magazines written in English. This corpus exceeds 8.7 billion words, collected from 2010 to the present time. I used this program to investigate how climate change is represented or framed in the media discourse by looking at the lexical terms (e.g., effects, Paris, impacts and combat) that are collocated with the climate change term. These terms belong to two parts of speech: nouns and verbs. Through looking at these parts of speech, I identified how climate change is represented in the world and U.S. media, where many metaphors are used for this representation in the media. Also, I used corpus analysis to investigate if the announcement sparked heated debate in the world and U.S. media.

O. K. Al-Shboul, *The Politics of Climate Change Metaphors in the U.S. Discourse*,
https://doi.org/10.1007/978-3-031-19016-2_8

Introduction

Corpus linguistics is defined as "the study of language based on examples of real-life language use" (McEnery & Wilson, 1996, cited in Baker, 2006, p. 1). This kind of study is conducted using technological tools. Sinclair (2005) also defines corpus as "a collection of pieces of language text in electronic form, selected according to external criteria to represent, as far as possible, a language or language variety as a source of data for linguistic research" (p. 16). Vaughan & O'keefe, 2016 state that

> Corpus linguistics involves the use of computers to rapidly search and analyze data bases of real language. These databases are called corpora (the plural of Latin corpus) and they can comprise any principled collection of written or transcribed spoken language. (p. 1)

Researchers (e.g., Taylor, 2007) maintain that there are disagreements over, or even no definitive answers to, whether corpus linguistics is a methodology, approach or paradigm. However, McEnery and Gabrielatos (2006) argue that corpus linguistics is one kind of empirical rather than theoretical linguistics.

Baker (2006) distinguishes corpus analysis from qualitative analysis by the fact that the former draws on more electronically encoded texts, and findings also are in the form of quantitative information. Because corpus analysis is quantitative in nature, the most prominent characteristics of this type of analysis is that it is based on the recurrences of entities such as lexical forms (Sinclair, 2005).

In addition to the qualitative analysis of the politicians' language, I used corpus analysis to investigate how climate change is represented or framed in the media discourse through looking at the lexical terms that are collocated with climate change. These terms belong to two parts of speech: nouns and verbs. Through looking at these parts of speech, I identified how climate change is represented in the world and U.S. media, where many metaphors are used for this representation in the media. The availability of electronic (digital) media has helped researchers investigate "how media texts might be repeatedly framing issues or events which are reported over a significant period of time" (O'Halloran, 2010, p. 356).

Also, I used corpus analysis to investigate if the announcement sparked heated debate in the world and U.S. media through looking at the frequency of the occurrences of the climate change/global warming terms and the Paris agreement/accord/deal terms before and after the President's announcement (of the withdrawal of the U.S. from the Paris climate agreement). Thus, to determine if this announcement sparked heated debate in the media, I applied two statistical tests: The Log-likelihood and Log Ratio to find whether the differences or gaps in the frequencies of the occurrences of these terms before and after the announcement are statistically significant and meaningful.

In this corpus analysis, I used the NOW program (News On the Web). Using this program (NOW), the researcher can refine the period of time he/she searches in (through using the *search by date* feature) and also which country he/she needs to search the data: United States, Canada, France, Australia, etc. For selecting the U.S. corpus, I refined the search through *sections* (where I selected the U.S. corpus). The NOW corpus comprises web-based newspapers and magazines written in English. This corpus exceeds 8.7 billion words, collected from 2010 to the present time. By way of explanation, I looked at the frequency of the occurrences of the climate change/global warming terms and the Paris climate agreement/accord/deal terms before and after the President's announcement (of the withdrawal of the U.S. from the Paris climate agreement). Thus, to determine if this announcement sparked heated debate in the media, I applied two statistical tests: The Log-likelihood and Log Ratio to find whether the differences or gaps in the frequencies of the occurrences of these terms before and after the announcement are statistically significant and meaningful.

Table 8.1 below shows the size of each corpus including the world and U.S. media one year before the announcement (05/31/2016–05/31/2017) and one year after the announcement (06/01/2017–06/01/2018); and also one month before the announcement (05/01/2017–05/31/2017) and one month after the announcement (06/01/2017–06/30/2017).

Corpus analysis. For this chapter, I used NOW software (News On the Web) where I drew on three corpus tools available in this program: *Collocate, Frequency and Compare*. I used this program since it involves data over the periods that I aimed to search in (one year and one month

Table 8.1 The corpus and its size

Name of corpus climate change	Total words in corpus
One year before announcement (the world media/full corpus)	1,691,232,855
One year after announcement (the world media/full corpus)	1,630,716,353
One month before announcement (the world media/full corpus)	154,054,062
One month after announcement (the world media/full corpus)	137,246,536
One year before announcement (US only)	181,811,078
One year after announcement (US only)	198,872,024
One month before announcement (US only)	15,676,012
One month after announcement (US only)	14,267,258

before and after the President's announcement of withdrawing the U.S. from the Paris climate agreement). The NOW corpus "contains 9.0 billion words of data from web-based newspapers and magazines from 2010 to the present time". They all written in English. These newspapers and magazines making up the corpus are from different countries including the United States, Britain, Canada, Ireland, Australia, New Zeeland, India, Sri Lanka, Pakistan, Bangladesh, Malaysia, Singapore, Philippines, Hong Kong, South Africa, Nigeria, Ghana, Kenya, Tanzania and Jamaica.

In this program (NOW), using the Collocate feature, I searched for the most common terms (nouns and verbs as specific parts of speech) that are most stringed with the climate change term. Accordingly, I searched for the most common nouns collocated with the climate change term to identify what are the aspects that the world and U.S. media mostly focuses on. Similarly, I searched for the verbs most collocated with the climate change term to identify what kinds of actions that the media emphasizes when discussing climate change.

Additionally, I used the *compare* feature available in the NOW software to find which is more frequently used: *developed* or *developing countries* terms in the media talking about climate change. Also, I compared the frequency of *people* to *animals* and *plants* in the media. This can give us statistical evidence that shows to what extent the media is concerned about living beings.

Moreover, as this book is about climate change and the Paris agreement, I used corpus analysis to investigate the frequency of these terms (climate change/global warming and Paris agreement/accords/deal) in one month and one year before and after the announcement (the decision of withdrawing the U.S. from the Paris agreement), as one way to see if this announcement sparked heated debate in the media. I, therefore, applied the Log-likelihood (LL) test to see if the differences in the frequencies of each of these terms before and after the announcement are statistically significant and I also applied the Log Ratio test to see if these differences are meaningful.

> Log Ratio is my attempt to suggest a better statistic for keywords/key tags than log-likelihood, which is the statistic normally used. The problem with this accepted procedure is that log-likelihood is a statistical significance measure—it tells us how much evidence we have for a difference between two corpora. However, it does not tell us how big / how important a given difference is. But we very often want to know how big a difference is! (Hardie, 2017, http://cass.lancs.ac.uk/author/andrewhardie/).

In sum, I drew on these three corpus tools: Collocate, Frequency and Compare for a systematic analysis of the media discourse on climate change and Paris agreement.

> Corpus linguists are thus able to make fairly confident generalizations about the varieties of language they are examining based on the combination of automated and human elements to the analysis. The automated side helps to direct the human researcher to aspects of the corpus that he or she may not have thought interesting to look at (Baker & Egbert, 2016, p. 2).

I used Collocate, Frequency and Compare tools available in this program (News On the Web) to answer these two main questions listed below. I investigated these questions in two separate sections. The first section answers the first question while the second section investigates the second question.

1. How is climate change represented in the world and U.S. media? What kinds of aspects and actions that the world and U.S. media

focuses on with regards to climate change? What are the most lexical units that co-occur with the climate change term? How frequent do the lexical items: people, animals and plants cooccur with the climate change term?
2. Did the President's decision of withdrawal from the Paris agreement spark heated debate in the media? (Did the climate change/global warming and Paris agreement/accord/deal terms increased in the world and U.S. media after the President's announcement to withdraw from the Paris climate agreement? Are the differences statistically significant?).

Section One: Aspects and Actions That the Media Focuses on About Climate Change

I used the Collocate tool available in the NOW program to investigate which lexical items, including nouns and verbs, that most frequently cooccur with the climate term so I can identify what aspects and actions that the world and U.S. media emphasizes when talking about climate change.

Before specifying parts of speech, I searched for the most common items stringed with the climate term in general (without specifying what parts of speech I need to search) to explore how climate change is predominantly represented in the world media, including the U.S. media. Figure 8.1 below shows the most five frequent lexical items (of all the parts of speech) that cooccur with the climate term over one year before the announcement (from 05/31/2016 to 05/31/2017 and one year after the announcement (from 06/01/2017 to 06/01/2018).

The first five common lexical items between the two periods are four nouns (effects, Paris, impacts and adaptation) and one verb (combat). For these nouns, they instantiate consequences related to climate change. For the verb (combat), it is metaphorically used (drawn from the source domain of war) with the climate term to emphasize the nature of this problem as being threatening, and correspondingly the necessity of responding to the consequences of this problem.

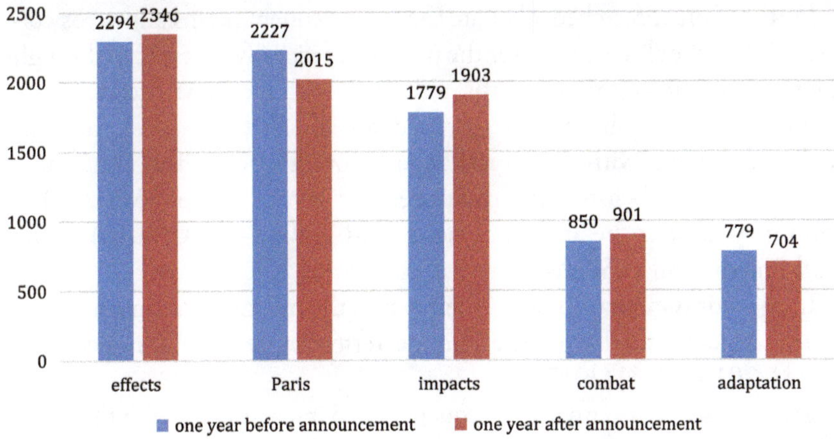

Fig. 8.1 The most frequent items collocated with the climate term in the world media

The first four items (effects, Paris, impacts and combat) occupy the same ranking in both periods (as the first, second, third and fourth positions before and after the announcement). We see that all lexical items that are collocated within climate change are concerned with the consequences (effects and impacts) and how to deal with the problem (adaptation and combat). Also, as the word *effects* is the first most common frequent item with the climate term in the media, it can be argued that the world media is concerned about this issue (climate change) mostly in terms of consequences. This is also supported by the high frequency of the lexical terms *impacts* and *combat* occupying the third and fourth positions, as illustrated in the Fig. 8.1 above. Based on the chart, we can see that the frequency of the word *Paris* is highest after the frequency of the word *effects*. This shows to what extent the Paris agreement was an important event all over the world. The word *Paris* was higher before the announcement. This might be attributed to the fact that the members signed the Paris climate agreement in Paris in 2015 and the agreement came into force in 2016. Consequently, the media worldwide shed light on that event that time. For the item *adaptation*, while it is the sixth word before the announcement, it occupies the fifth position after the announcement. Accordingly, the word adaptation is the fifth most frequent word shared in both periods.

Furthermore, I searched what are the most frequent nouns and verbs collocated with the climate term for the purpose of identifying aspects (through nouns) and actions (through verbs) associated with the climate term.

For the noun collocates, I searched for the first five most frequent nouns associated with climate change before and after the announcement: one year before the announcement (05/31/2016–05/31/2017) and one year after the announcement (06/01/2017–06/01/2018) in the world media and U.S. media.

In the world media, Table 8.2 below lists the most five common nouns which are collocated with the climate term before the announcement (05/31/2016–05/31/2017).

These items all instantiate consequences and impacts to do with this environmental problem (climate change). Besides, the lexical term *agreement* is among the most frequent terms not only before the announcement but also after the announcement. This can be a concrete piece of evidence that the world media discussed this agreement (the Paris climate agreement) very largely even before the announcement. And also, for the most five common nouns in one year after the announcement (06/01/2017–06/01/2018) in the world media they are listed in Table 8.3.

Based on the corpus findings, it can be clearly noticed that the world media sheds light on climate change in terms of its consequences. This is

Table 8.2 The most five common nouns collocated with the climate term before the announcement in the world media

Order	Lexical items (nouns)	Frequency
1	EFFECTS	2292
2	AGREEMENT	2029
3	ENVIRONMENT	1987
4	IMPACTS	1779
5	IMPACT	1468

Table 8.3 The most five common nouns collocated with the climate term after the announcement in the world media

Order	Lexical items (nouns)	Frequency
1	EFFECTS	2346
2	ENVIRONMENT	2290
3	IMPACTS	1903
4	IMPACT	1554
5	AGREEMENT	1538

evidenced through the most frequent items in both of the periods (before and after the announcement) that cooccur with the climate term: *effects and impact(s)*. These lexical terms communicate consequences of this problem. This conveys the concerns that the world media has about this problem (climate change). Furthermore, the word *agreement* is among the most frequent items used before and after the announcement. This can serve as statistical evidence that the media sees the Paris climate agreement an important event.

Similarly, the corpus analysis of the U.S. media shows, to a large extent, similar findings. That is, the most frequent common nouns stringed with the climate term in the U.S. media before and after the announcement are *effects, science, and agreement* as shown in the two Tables 8.4 and 8.5.

For the most frequent nouns collocated with the climate term after the announcement in the U.S. media, as listed in the Table 8.5 below.

It can be seen that in the world media and U.S. media, there are common items stringed with the climate term. These common items are *effects*, *impacts* and *agreement*. This statistical evidence to a large extent reveals that the U.S. media and world media express and share the same concerns about climate change with regard to its consequences (as demonstrated by the most common nouns collocated with the climate term as shown above). Besides, the term *agreement* is among the most collocated items with the climate term. This demonstrates that the Paris

Table 8.4 The first five frequent nouns collocated with the climate term before the announcement in the U.S. media

Order	Lexical items (nouns)	Frequency
1	EFFECTS	211
2	SCIENCE	142
3	TRUMP	142
4	AGREEMENT	136
5	ISSUES	136

Table 8.5 The first five frequent nouns collocated with the climate term after the announcement in the U.S. media

Order	Lexical items (nouns)	Frequency
1	EFFECTS	258
2	IMPACTS	154
3	IMPACT	129
4	SCIENCE	115
5	AGREEMENT	102

climate agreement represents an important event in many countries around the world. One reason for this is that it is viewed by politicians (e.g., governors and mayors in the U.S.) as an effective solution to tackle climate change and an important way to mitigate its effects. This explains why the media brings the agreement into light. In 2016, one nighty five countries signed this agreement in Paris. This event heavily attracted the media to talk about it. This also shows that this agreement is strongly present in the media especially when it comes to talk about the consequences, instantiated by such lexical terms as *effects and impact(s)*.

Turning to the next part of speech, which is *verbs* to investigate (what are) actions that are most emphasized and discussed with regard to climate change in the media.

Table 8.6 below shows the first five verbs which are collocated with the climate term in one year before the announcement (05/31/2016–05/31/2017) in the world media.

For the first five verbs collocated with the climate term in one year after the announcement (06/01/2017–06/01/2018) in the world media are listed in the Table 8.7 below.

Accordingly, in the world media, over the two periods (before and after the announcement), it can be clearly seen that the most frequent lexical verbs that cooccur with the climate term are the same: *combat, fight, tackle,* as indicated in Tables 8.6 and 8.7 above. These verbs (metaphorical expressions) are drawn from the source domain of war. They are metaphorically used to convey the meaning of responding to climate change promptly. They thus stress urgency.

Similarly, in the U.S. media, the most frequent verbs stringed with the climate term communicate urgency to respond to climate change. The lexical verbs *combat* and *fight* are metaphorically used with the climate

Table 8.6 The first five frequent verbs collocated with the climate term before the announcement in the world media

Order	Lexical items (verbs)	Frequency
1	COMBAT	808
2	ADDRESS	792
3	FIGHT	707
4	TACKLE	602
5	ADDRES	537

Table 8.7 The First five frequent verbs collocated with the climate term after the announcement in the world media

Order	Lexical items (verbs)	Frequency
1	ADDRESS	875
2	COMBAT	866
3	TACKLE	681
4	FIGHT	676
5	MITIGATE	528

Table 8.8 The first five frequent verbs collocated with the climate term before the announcement in the U.S. media

Order	Lexical items (verbs)	Frequency
1	COMBAT	118
2	BELIEVE	95
3	FIGHT	87
4	ADDRESS	75
5	ADDRESSING	72

Table 8.9 The first five frequent verbs collocated with the climate term after the announcement in the U.S. media

Order	Lexical items (verbs)	Frequency
1	ADDRESS	105
2	COMBAT	91
3	RISING	72
4	FIGHT	64
5	BELIEVE	60

term in the U.S. media as shown in the Tables 8.8 and 8.9 on the next page from 05/31/2016 to 05/31/2017 before the announcement and 06/01/2017 to 06/01/2018 after the announcement.

In conclusion, the world media and the U.S. media highlight climate change in terms of the consequences and how to deal with this problem, which is evidenced through the most frequent nouns collocated with the climate term which convey the consequences of this problem (such as *effects and impacts*). The verbs, in turn, *combat, fight* and *tackle* stringed with the *climate* term are used to communicate responding with a sense of urgency to these consequences and concerns. They are drawn from the source domain of war. The occurrences of these items as collocates with the climate term indicate that the world is suffering from serious consequences and the problem is getting worse. The media thus throws light on the importance of initiating measures and procedures to deal with this problem. Accordingly, climate change is framed as a threatening problem that requires immediate responses to tackle the problem. In addition, the lexical verb *believe* is among items most stringed with climate change in the U.S. media. This can indicate that the media in the U.S. discusses the belief about climate change. For example, whether combating climate change should be a top priority or not or also whether it is attributed to human activity or not.

As this book is about climate change and its consequences on the environment (including people, animals and plants), it is worth investigating the frequency of occurrences of the *people, animals, and plants* terms as collocates with the climate term in the media discourse, using the compare tool. This is one way to know to what extent the media sheds light on these elements of the environment including *people, animals and plants* in comparison with one another when talking about climate change, and thereby indicating to what extent the media is concerned about each of these elements in comparison with one another. There is no doubt that people, animals and plants are important words associated with environmental problems like climate change as being the most affected elements suffering from its consequences. I used corpus analysis to investigate if there are big differences with regards to the frequency of the occurrences of these terms. The frequency of these terms can tell us which one of these elements that the media express more concern than the others. Using corpus analysis, it can be clearly seen that the lexical item *people* co-occurs far more frequently with the climate term than the lexical item *animals or plants*. While the frequency of occurrences of people is (2769), the frequency of occurrences of the term *animal(s)* is (336) or *plant(s)* is (316) as listed in the Table 8.10 below. The media uses more other terms than, to refer to, *people, animals and plants*. However, the results of the frequency

Table 8.10 The frequency of the occurrences of people, animals and plants

Order	Lexical items	Frequency
1	People	2769
2	Animal(s)	336
3	Plant(s)	316

of occurrences of these general terms (people, animals and plants) can serve as concrete evidence that the media is far more considerate and concerned about people's lives than *animals'* or *plants'*.

Besides to searching for these lexical terms (people, animals and plants) I searched for the frequency of the occurrences of developing and developed countries terms using the compare tool. That is, since countries are seen as agents (e.g., polluters) and also as affected members by the media, it is important to use corpus analysis to investigate how frequent developed and developing countries are used in the media. Among the questions which I investigated how frequently *developed country(ies)* term is used as a collocate with the climate term in comparison with *developing country(ies)* term. Using corpus analysis, I found that developing country(ies) term (714 occurrences) is used much more than the *developed country(ies)* term (163 occurrences) in the worldwide media as listed in Table 8.11 below.

Developing countries take bigger space in the media as is evidenced in the higher frequency of the occurrences of *developing countries* term than that of the *developed countries* term. In this regard, a report in 2019 from the government of Canada demonstrates that developing countries are more vulnerable to climate change:

> Developing countries are the most impacted by climate change and the least able to afford its consequences. Their vulnerability is due to multiple factors that can limit their ability to prevent and respond to the impacts of climate change. Climate change has the potential to reverse significant development gains made in these countries. According to the World Health Organization, as of the year 2030, climate change is expected to contribute to approximately 250,000 additional deaths per year, from malnutrition, malaria, diarrhea and heat stress. (The government of Canada website, 2019) (https://www.international.gc.ca/worldmonde/issues_developmentenjeux_developpement/environmental_protectionprotection_environnent/climate-climatiques.aspx?lang=eng)

Table 8.11 The frequency of developed and developing countries collocated with the climate term

	Classifications of countries	Frequency
1	Developed countries	163
2	Developing countries	714

This can explain why the media sheds light on developing countries (as being the most affected regions in the world). Another reason of why the media sheds light on developing countries is the interest of developed countries to involve the developing countries in addressing climate change since it is a world phenomenon so that all countries need to take part in.

Section Two: Debate on Climate Change in the Media

As this book investigates the texts produced by the governors and mayors after the President's announcement of withdrawing from the Paris agreement. I carried out a further corpus analysis to investigate if the media devoted much attention to this decision (and correspondingly if the announcement has an impact on the frequency of the occurrences of these terms). I thus looked at the frequency of occurrences of particular terms including climate change/global warming and Paris agreement/accord/deal terms over specific periods of time before and after the announcement. I also applied two statistical tests (Log-likelihood and Log Ratio) to see whether the differences between the frequency of occurrences before and after the announcement are statistically significant and if they are meaningful. That is, a Log-likelihood result can be significant but not very meaningful if the Log Ratio result is less than 0.5%, (where 0.5% represents a 50% increase in the use of the term, suggesting a meaningful, noticeable change in frequency), especially in a corpus of this size. These statistical tests are one way to tell us if the President's decision of withdrawing the U.S. from the Paris agreement sparked heated debate in the media (through looking at the differences of frequencies of occurrences of these terms before and after the announcement).

Figure 8.2 shows that while the frequencies of occurrences of the *climate change* and *global warming* terms in the world media before the announcement in one year are (69243) and (10683), the occurrences of these two terms after the announcement in one year are (68668) or (9568) respectively.

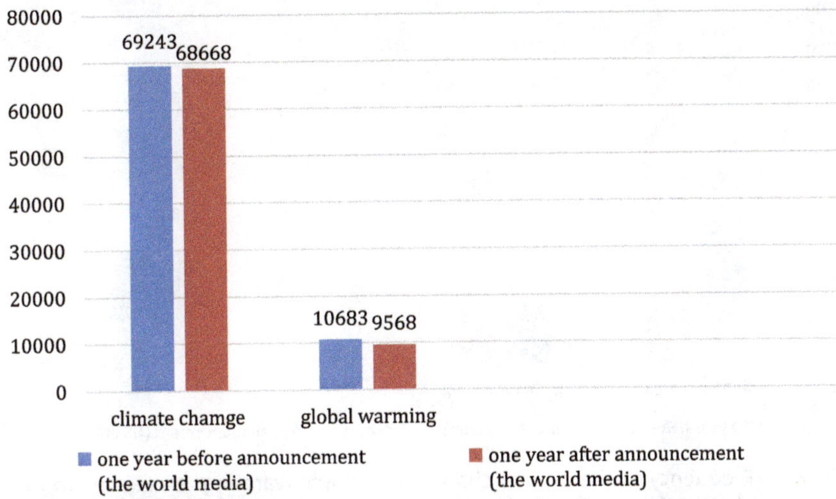

Fig. 8.2 Frequency of the climate change and global warming terms before and after the announcement in the world media

I applied the Log-likelihood and Log Ratio tests, the Log-likelihood for the frequency of occurrences of the climate change term is 27.21 and the Log Ratio of the frequencies of occurrences between the two corpora is 0.04. Similarly, the Log-likelihood for the global warming term is 27.51 and the Log Ratio of this term is 0.11. Accordingly, for the climate change term, although the result of the Log-likelihood indicates that the difference is statistically significant, the Log Ratio result of 0.04 suggests that this difference is very small and may not be meaningful. The findings for the global warming term are similar, where the Log-likelihood of 27.51 indicating that the difference is statistically significant. However, the Log Ratio of 0.11 suggests that this difference is not meaningful.

The corpus analysis of the U.S. media coverage shows similar patterns to that of the world media with regards to the size of the gaps in the frequencies of occurrences of the climate change and global warming terms in one year before and after the announcement. The frequency of occurrences of *climate change* and *global warming* terms are 7918 and 1574 before the announcement and 7506 and 1424 after the announcement respectively, as indicated in the Fig. 8.3 below.

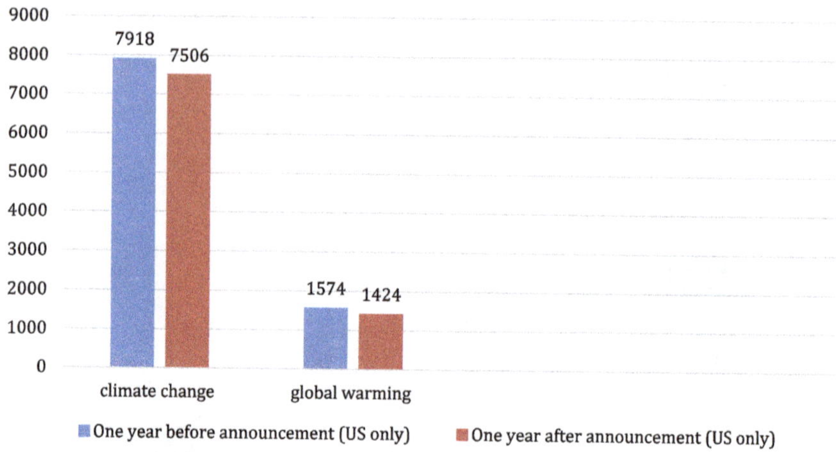

Fig. 8.3 Frequency of the climate change and global warming terms one month before and after the announcement in the U.S. media

Applying the statistical tests, I found that the Log-likelihood for the climate change term in the U.S. corpus is 78.97 and the Log Ratio is 0.21. Also, the Log-likelihood of the global warming term is 26.99 and the Log Ratio is 0.27. Based on the Log-likelihood results, this test shows that there are statistically significant differences between the gaps for each of the terms (climate change and global warming) before and after the announcement. However, the differences are not big or meaningful as the Log Ratio results 0.21 and 0.27 for both of the terms (climate change and global warming) are less than 0.5, with both of which more frequent in the corpus prior to the announcement.

But to investigate more closely if the announcement has an impact on the frequency of the occurrences of *climate change* or *global warming* terms (so we can be more able to determine if this announcement sparked heated debate), it is worth looking at the frequencies of the occurrences of these items in closer periods to the date of the announcement in the world media and the U.S. media, which is on a month basis: one month before the announcement (from 05/01/2017 to 05/31/2017) and one month after the announcement (from 06/01/2017 to 06/30/2017).

For the world media, Fig. 8.4 below shows the frequencies of these terms in one month before and after the announcement:

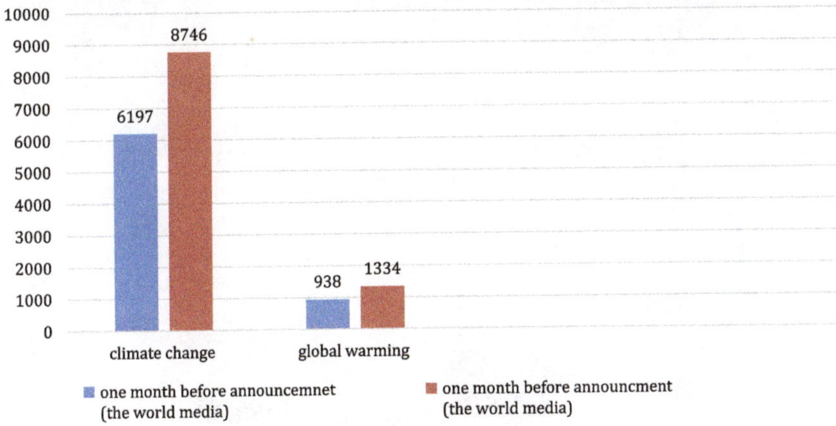

Fig. 8.4 Frequency of the climate change and global warming terms one month before and after the announcement in the world media

The frequency of occurrences of *climate change* and *global warming* terms over one month before the announcement are 6197 and 938; and the frequency of occurrences of these terms over the same period after the announcement are 8746 and 1334 respectively.

I applied the statistical tests to find out whether the differences in the frequencies of occurrences of these terms before and after the announcement are statistically significant and if they are meaningful: The Log-likelihood for the *climate change* term is 781.25 and the Log Ratio is 0.66. For the *global warming*, the Log-likelihood is 122.70 and the Log Ratio is 0.67. Accordingly, the Log-likelihood results indicate that the differences for both these terms are statistically significant. Similarly, the findings of the Log Ratio 0.66 and 0.67 for *climate change* and *global warming* terms respectively suggest that the gaps in the frequencies are meaningful with both terms more used after the announcement than prior to the announcement.

The Fig. 8.5 above gives information about the frequency of these terms one month before the announcement (from 05/01/2017 to 05/31/2017) and after the announcement (from 06/01/2017 to 06/30/2017) in the U.S. media. The occurrences of the *climate change* term duplicated from 594 to 1072 and also that of the *global warming* term from 98 to 202 over this period. The gaps of frequencies of these

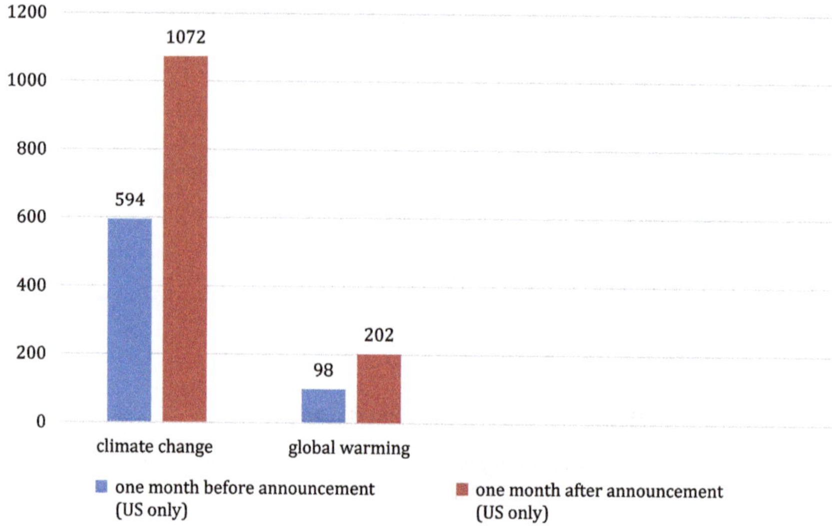

Fig. 8.5 Frequency of the climate change and global warming terms one month before and after the announcement in the U.S. media

terms in the U.S. media are bigger than that of the world media over the same period. Applying the statistical tests, the result of the Log-likelihood for the gaps of the climate change term is 187.79. For the *global warming* term, the result of the Log-likelihood is 47.27. The results suggest that the differences are statistically significant. Also, the Log Ratio for both the climate change and global warming terms are 1.18 and 0.99 respectively. Based on the Log Ratio results, the differences are very meaningful, with both terms used twice as much after the announcement than before the announcement.

Accordingly, based on the results of Log Ratio, the differences of frequencies are bigger when these terms are compared on a month basis (one month before and after the announcement) than on a year basis (one year before and after the announcement) in both the world media and U.S. media. The bigger gaps in the frequencies of occurrences of the *climate change* and *global warming* terms on the month basis is attributed to the President's announcement of withdrawing the U.S. from the Paris agreement (about climate change) on June 1 2017. This announcement increased the frequency of occurrences of the *climate change* and *global warming* terms after the announcement in the world and U.S. media, making the gaps bigger on one-month to one-month basis than one-year to one-year basis (before and after the announcement).

This leads us to investigate the frequency of the Paris agreement/accord/deal terms over the same periods. The Paris agreement represents an important event linked with this problem (climate change). It was signed in Paris by many countries around the world. Based on reviewing many news articles, the agreement was referred to by several terms including *Paris agreement, Paris accord and Paris deal.* Hence, I searched for all these terms. I searched the frequency of these terms in the same periods as that of the climate change and global warming terms: on a year and month basis (one year before and after the announcement, and one month before and after the announcement).

In the world media, according to Fig. 8.6, the frequencies of occurrences of the *Paris agreement*, *Paris accord* and *Paris deal* terms are 7590, 652 and 443 respectively in one year before the announcement and are 7774, 2075 and 559 respectively in one year after the announcement.

The statistical test shows that the Log-likelihood for the *Paris agreement* term is 14.01 and the Log Ratio is 0.09. This shows that while the

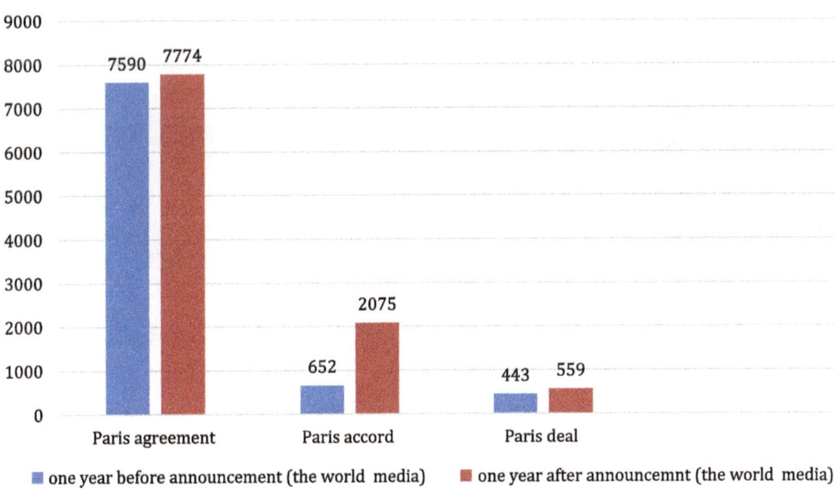

Fig. 8.6 Frequency of the Paris agreement, Paris accord and Paris deal terms one year before and after the announcement in the world media

difference is statistically significant according to the Log-likelihood, the Log Ratio result suggests that the gap is not meaningful. For the *Paris accord*, the Log-likelihood is 833.32 and the Log Ratio is 1.72. The finding of the Log-likelihood suggests that the gap is statistically significant and that of the Log Ratio indicates that the gap is meaningful, with the *Paris accord* term used more three times as frequently after the announcement than prior to the announcement. With regards to the *Paris deal* term, the Log-likelihood is 18.02 and the Log Ratio is 0.39. The Log-likelihood result suggests that the difference is statistically significant, but the Log Ratio indicates that this difference is not meaningful.

Moving to the U.S. media, Fig. 8.7 below shows the differences between the frequencies of occurrences of those terms (the Paris agreement, Paris accord and Paris deal terms) in one year before and after the announcement.

As Fig. 8.7 above illustrates, the gaps are bigger in the U.S. media than that of the world media before and after the announcement. The frequencies of the occurrences increased from 490 to 729 for the *Paris agreement* term; from 90 to 326 for the *Paris accord* term and from 37 to 70 for the *Paris deal* term over the period from 05/31/2016 to 05/31/2017 and from 06/01/2017 to 06/01/2018.

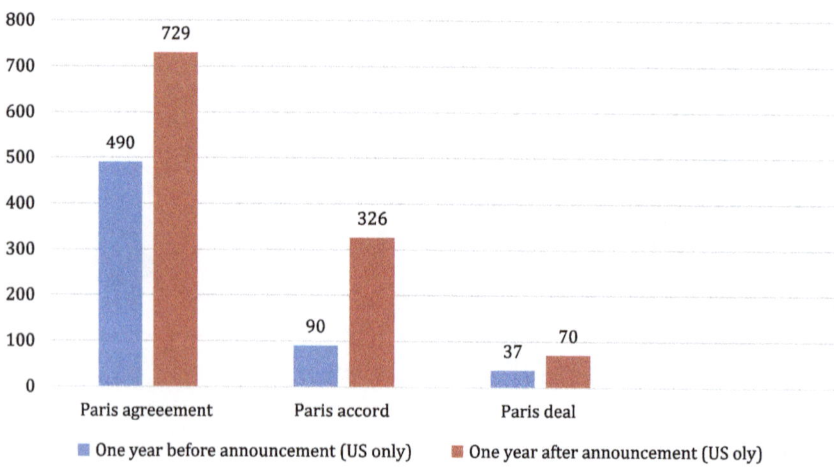

Fig. 8.7 Frequency of the Paris agreement, Paris accord and Paris deal terms one year before and after the announcement in the U.S. media

Applying the statistical tests, the Log-likelihood for the *Paris agreement* is 28.18 and the Log Ratio is 0.44. For the *Paris accord*, the Log-likelihood is 121.86 and the Log Ratio is 1.73. As for the *Paris deal* term, the Log-likelihood is 7.60 and Log Ratio is 0.79. Based on results of the Log-likelihood for each of these terms, the differences of the frequencies of occurrences of each term are statistically significant. The results of Log Ratio suggest that the gaps for all the terms are meaningful except for that of the *Paris agreement* term 0.44. It is, however, close to 0.5, which represents a 50% increase. Based on the Log Ratio results, the gaps of the differences of the occurrences of the *Paris agreement*, *Paris accord* and *Paris deal* in the U.S. media are bigger and more meaningful than that of the world media over the year basis.

Figure 8.8 below illustrates the frequencies of the *Paris agreement*, *Paris accord* and *Paris deal* terms over one month (before and after the announcement) in the world media: 938, 219 and 118 before the announcement and 2968, 1141 and 276 after the announcement respectively.

Applying the statistical test, I found that the Log-likelihood for the *Paris agreement* term is 1356.08 and the Log Ratio is 1.83, for the *Paris*

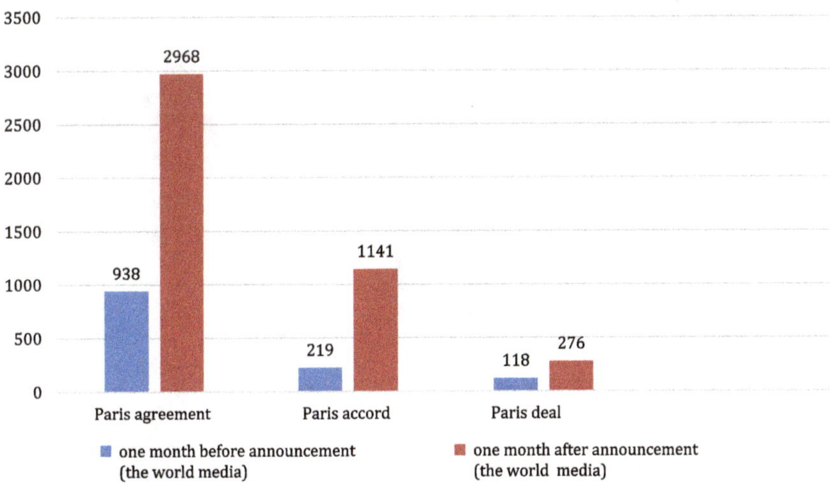

Fig. 8.8 Frequency of the Paris agreement, Paris accord and Paris deal terms one month before and after the announcement in the world media

accord term the Log-likelihood is 795.88 and the Log Ratio is 2.55 and for the *Paris deal* term the Log-likelihood is 84.75 and the Log Ratio is 1.39. Accordingly, based on the Log-likelihood, the differences in the frequencies of occurrences of each of these three terms before and after the announcement are statistically significant. Also, the Log Ratio results demonstrate that the gaps in the frequencies of occurrences of each of these three terms suggest a large effect size, and correspondingly meaningful, with these three terms used two to three times as frequently after the announcement as before the announcement.

With regards to the U.S. media, Fig. 8.9 below gives information for the frequency of these terms (Paris agreement/accord/deal) over the same periods. It can be clearly seen that the gaps in the frequencies of these terms in the U.S. are even bigger than that in the world media when compared on a month basis.

While the term *Paris agreement* term occurred 75 in one month before the announcement in the U.S. media, this figure increased about five times 379 in one month after the announcement. Similarly, the term *Paris accord* increased about six times from only 34 before the announcement in one month to 198 in one month after announcement. Also, the

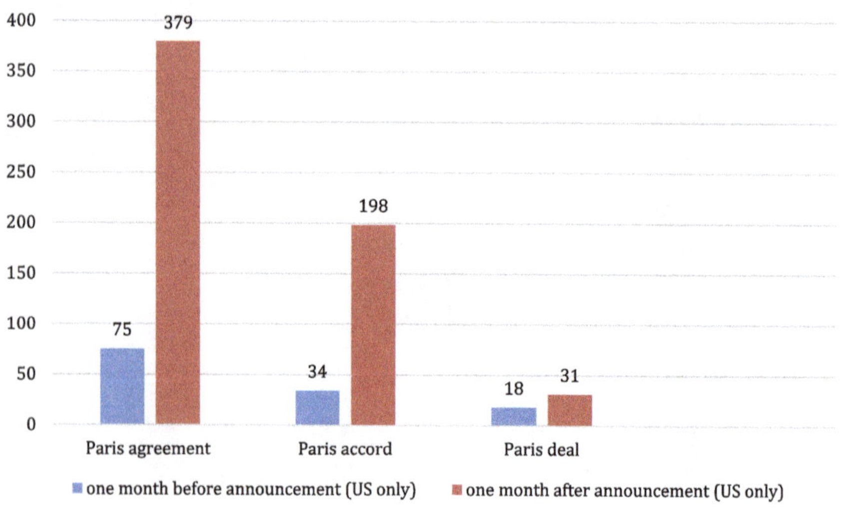

Fig. 8.9 Frequency of the Paris agreement, Paris accord and Paris deal terms one month before and after the announcement in the U.S media

occurrences of *Paris deal* term doubled from 18 to 31 time. The statistical test shows that the Log-likelihood for the Paris agreement term is 252.05 and the Log Ratio is 2.47; for the Paris accord term the Log-likelihood is 144.24 and the Log Ratio is 2.68. For each of these, the Log Ratio values indicate that the terms were used around six times more frequently following the announcement. For the Paris deal term, the Log-likelihood is 4.82 and the Log Ratio is 0.92, indicating that it is used nearly twice as much following the announcement. All the scores of the Log-likelihood suggest that the differences in the frequencies of occurrences of these terms before and after the announcement are statistically significant. Furthermore. The findings of the Log Ratio of these terms indicate that these gaps are very meaningful. Based on the results of the Log Ratio, the gaps in the frequencies of occurrences of these terms both in the world and U.S media in one month before and after the announcement are very big and meaningful. This can give us concrete evidence that the decision of withdrawal from the Paris climate agreement sparked heated debate in the media particularly in the U.S. media.

To sum up, the figures above (in this section) show that the gaps in the frequencies of occurrences of the *Paris agreement, Paris accord and Paris deal* terms are bigger than that of the *climate change* and *global warming* terms before and after the announcement (both in one-month and in one-year basis). This indicates that the media heavily reacted to the President's statement on the withdrawal from this (Paris) agreement. In his speech I analyzed; the President mentioned the agreement very frequently. He referred to the agreement twenty-eight times. He referred, in turn, to the climate only one time, to make a claim that the agreement works for the interest of other countries but not the United States: "This agreement is less about *the climate* and more about other countries gaining a financial advantage over the United States. The rest of the world applauded when we signed the Paris Agreement—they went wild; they were so happy". In his statement, the President focused on the deal itself rather than the environmental problem (climate change) and its consequence. He justified his decision by the claim that the Paris agreement is not fair and undermines the economy of the country. Withdrawing, therefore, is a necessary action. Thus, it is not surprising that the media shed light on this agreement after his speech, where he used metaphors to conceptualize the agreement as unfair and confiner. This explains the big differences between the occurrences of the *Paris agreement, Paris accord* and *Paris deal* terms before and after the announcement in the media in one month before and after the period. This heated debate over the Paris agreement, after the official announcement, is one important piece of evidence that the media both in the U.S. and worldwide is very concerned about climate change since this decision (withdrawing the U.S. from the Paris agreement) can make combating climate change more difficult. This can be supported by the statistical findings in the first section of this chapter that the most collocated lexical items with the *climate change* term in the media are *effects* and *impacts* and also verbs which are drawn from the source domain of war: *combat, fight* and *tackle.*

Conclusion

The corpus analysis broadened the scope of this book by investigating the media so that we can have a bigger picture of how climate change is framed by looking at how this problem is framed not only in the U.S. politics but also in the media. This chapter thus aimed to investigate two important questions using corpus tools: how climate change is represented in the media? and if the media devoted a large amount of attention to this decision (the president's decision of withdrawing the U.S. from the Paris climate agreement).

For the first section of this chapter investigating how climate change is conceptualized, I looked at the most frequent items collocated with the *climate change* term as one way to see how the media describes and frames climate change. Based on these items, this problem (climate change) is conceptualized as an immediate threat. That is, the world and U.S. media emphasizes responding to this issue urgently. This can be indicated through using verbs, cooccurring most frequently with the *climate change* term. These lexical items (verbs) are drawn from the source domain of War (e.g., combat and fight). In addition, lexical items such as *effects* and *impacts* are among the most collocated nouns with the *climate change* term. This also reveals to a large extent that the media focuses on and discusses the disastrous consequences of this problem. Thus, framing climate change as a threat is what the politicians and the media emphasize (since they very frequently drew on the source domain of war). These terms of War were metaphorically used by the politicians (the governors and mayors) as well as the media (the world and U.S. media). Accordingly, the source domain of War is the common domain which the politicians and the media draw on to talk about climate change.

With regards to the second section investigating the reaction of the media to the decision of withdrawal from the Paris agreement and if the media devoted a large amount of attention to this decision, I looked at the frequency of occurrences of particular terms: *climate change/ global warming and Paris agreement/ accord/deal* terms before and after the announcement. I applied two statistical tests: The Log-likelihood and Log Ratio to see whether the differences are statistically significant and meaningful. The findings of the statistical tests suggest that the media devoted much attention to this decision, particularly the U.S. media since the results of the statistical tests for the differences of frequencies of the occurrences of the *climate change/ global warming and Paris agreement/ accord/deal* terms in the U.S. media are greater than that of the world media. Besides, the statistical results indicate that the gaps of the frequencies of the occurrences of the *Paris agreement/accord/deal* terms are bigger than those of the *climate change/global warming* terms before and after the announcement. This suggests that the media devoted a larger amount of attention to this agreement than the problem itself (climate change) in that period of time. This is attributed to the fact that the President's statement (as discussed in the previous chapter) focused on the Paris agreement rather than climate change and/or how to tackle this problem. Consequently, the media focused much more on this agreement after the announcement than any other time ever.

References

Baker, P. (2006). *Using Corpora in Discourse Analysis*. Continuum.

Baker, P., & Egbert, J. (2016). Introduction. In P. Baker & J. Egbert (Eds.), *Triangulating Methodological Approaches in Corpus-Linguistics Research* (pp. 1–19). Routledge.

Hardie, A. (2017). CASS in the 2017 ESRC Festival of Social Science. ESRC Centre for Corpus Approaches to Social Science (CASS). http://cass.lancs.ac.uk/author/andrewhardie/

McEnery, T., & Gabrielatos, C. (2006). English Corpus Linguistics. In B. Aarts & A. McMahon (Eds.), *The Handbook of English Linguistics* (pp. 33–71). Blackwell.

O'Halloran, K. (2010). How to Use Corpus Linguistics in the Study of Media Discourse. In A. O'Keeffe & M. McCarthy (Eds.), *The Routledge Handbook of Corpus Linguistics* (pp. 563–577). Routledge.

Sinclair, J. (2005). Corpus and Text: Basic Principles. In M. Wynne (Ed.), *Developing Linguistic Corpora: A Guide to Good Practice* (pp. 1–16). Oxbow Books.

Taylor, C. (2007). What Is Corpus Linguistics? What the Data Says. *ICAME, 3*, 179–200.

The White House. Our Government. https://www.whitehouse.gov/about-the-white-house/our-government/

Vaughan, E., & O'keefe, A. (2016). Corpus Analysis. *The International Encyclopedia of Language and Social Interaction*, 1–17.

9

Ecosophy

In the ninth chapter, I wrote my ecological philosophy. Ecosophy (ecological philosophy) constitutes an important theme in ecolinguistic studies. This is one important way to extend for ecolinguistics and make it clearer for the reader how this book contributes to ecolinguistics. It (ecosophy) is the basis upon which ideologies and linguistic expressions bearing those ideologies are judged as positive or negative (Arne Naess, 1989). Through my ecosophy, I can reveal this position that I am taking towards the problem that I am investigating. In this chapter I also discussed the politicians' perspectivization according to this ecosophy.

Ecological Philosophy (Ecosophy)

In this chapter, I wrote my ecological philosophy (ecosophy). But before that, I explained the relationship between ecosophy and ecolinguistics. Ecosophy is the explicit commitment toward the environmental issue which I am investigating, which is climate change in this context. It (ecosophy) is the basis upon which ideologies and linguistic expressions bearing those ideologies are judged as positive or negative. The term ecosophy first outlined and introduced in Arne Naess' works (1989). Naess proposed his own ecosophy as a model, where he emphasized the

O. K. Al-Shboul, *The Politics of Climate Change Metaphors in the U.S. Discourse*,
https://doi.org/10.1007/978-3-031-19016-2_9

intrinsic value of nature. He formulated his own ecosophy in which he refuted that living beings are ranked based on their relative in-trinsic value. He emphasized that animals and plants have no less right to live and to be respected than humans.

Stibbe defines Ecosophy as "an ecological philosophy, i.e. a normative set of principles and assumptions about relationships among humans, other forms of life and the physical environment. Analysts use their own ecosophy to judge the stories that they reveal through linguistics analysis" (2015, p. 202). Thus, "once analysts have determined their own ecosophies, they can use them to provide criteria for judging whether discourses are positive or not" (Stibbe, 2018, p. 175). Ecosophy, which is one contribution to Ecolinguistics, is CDA oriented, where it (CDA) demands an overt commitment on the part of the CDA analyst. This requires that the analyst reveals which side (group) he/she supports (Carvalho, 2018; Fairclough & Wodak, 1997). Therefore they "…want to understand, expose, and ultimately challenge social inequality" (Van Dijk, 2015, p. 466). Accordingly, those metaphors that I considered as positive or negative may not be so for another analyst or researcher. This, to a large extent, depends on what philosophy that he/she adopts towards the environmental problem (e.g., is I am with those who support or denounce the exit).

CDA provides explanations and critique on the part of the researcher (Gee, 2014; Sum & Jessop, 2013). This means that the researcher's attitude determines the explicit position that he/ she takes. Through my ecosophy, I can reveal this position that I am taking towards the problem that I am investigating. As this book investigates texts on environmental issues from the CDA perspective, this analysis should be based on ecolinguistics, which is concerned with environmental discourse.

To begin, climate change is real, and it is a threat to the globe we live in. Although political consensus does not exist till this time, scientific consensus emphasizes its dangerous impacts on the environment. Thus, it is not confined to one area or region, but it is (and should be) of concern to all the countries around the world. We thus need to act on this problem to protect the environment as the environment has an intrinsic value. I use here the term environment to include all living beings since I believe not only humans but also animals and plants to be respected and have the right to have an environment in which all living beings can live safely. Accordingly, profits should not be gained at the expense of damaging the environment. The Paris agreement, therefore, represents an important and serious effort that most if not all countries around the world signed and committed to deal with climate change before the problem is getting even worse. Therefore, it is a collective responsibility. This agreement aims to ensure that its members pledges committing to its terms and that they effectively apply these commitments. However, it is not only the governments' responsibilities, but it is a responsibility of individuals. Responsibility falls on the government as well as on individual levels. Stibbe maintains that "although ecosophies are fundamentally a statement of values and assumptions, they also need to be based on evidence, and adapted as further evidence emerges" (2017, p. 15). In this regard, evidence is strongly present. What most countries are suffering from due to climate change are considered a strong piece of evidence that climate change occurs, and it is human-made. There are many serious consequences for this problem. By way of example, what is happening now in the natural world including floods, rise in temperature is attributed, to a considerable extent, to changes in climate. I do believe that climate change is human-made, and consequently if we do not stop our behavior of continuing relying on industries with high emissions, then we will have been suffering from more and more consequences. Now we see more and more companies and factories tend to invent and design machines working on clean (wind and solar) energy. This is considered to be a movement in the right direction. However, conventional energy is still predominantly used inside and outside the United States. Activists and politicians should work harder to raise people's awareness of the dangers of climate change by educating them the causes and consequences.

Classification of the Politicians' Perspectivization Based on My Ecosophy

Perspectivization is "positioning speaker's or writer's point of view and expressing involvement or distance" (Reisigl & Wodak, 2008, p. 95). Numerous researchers like Charteris-Black (2014, 2018) and Fløttum

(2017) emphasize that metaphor is among the most important devices that can be used for the speaker's/writer's perspectivization. As discussed in the sections above, perspectivization or framing is performed through the discourse of (de)legitimization through which the politicians aim to transform their attitude into a reality, and thereby should be accepted and adopted, to then trigger social change.

Van Dijk classified ideologies as *positive* and *negative* (Van Dijk, 1997, 2015). Correspondingly, Stibbe (2015) argues that ideologies can be evaluated as *positive* or *negative* based on the analyst's ecosophy. That is, if they support and resonate with the analyst's ecosophy, then they can be classified as positive ideologies. But if not, then they are negative ideologies. Accordingly, in the current study, I follow this (Stibbe's) classification of ideologies. That is, based on my ecosophy (the analyst's ecosophy), the politicians' ideologies in the study are either positive or negative. Therefore, the governors' senators' and mayors' attitude (opposing the exit) towards climate change, on the one hand, represents the positive ideology. The president's and senators' attitude (supporting the exit) towards climate change, on the other hand, is the negative ideology. The Table 9.1 indicates this classification: The president, governors and mayors represent as two main groups in the study, with each holding a different attitude.

In the light of the data analyzed in this book, the politicians in each group, reacting against the decision, have two main different but related challenges: the first is to make their audience perceive their attitude as a reality (so they are invited to be involved in this group to adopt this

Table 9.1 The participants' classifications

The participants	Classification of participants	Perspectivization attitude	Evaluated by the researcher
The President and other politicians supporting exit	First group	Paris agreement is not good (Climate change is not a top priority)	Negative ideology or Positive ideology
Governors Senators and Mayors opposing exit	Second group	Paris agreement is good (Climate change is a top priority)	

reality), and the other challenge in turn is to convince the audience that the point of view of the other side (the oppositional attitude) is not a reality but merely a negative ideology, and therefore needs to be ignored (and so they need to be distant from this ideology). "What is knowledge for one group may be seen as an ideology by others" (Van Dijk, 1997, p. 28). Thus, the reality which the governors and mayors that they try to construct is that climate change is open end, dangerous, causes pollution, and therefore governments and people need to work hard to deal with this threat (climate change).

> Ideologies → (discursive practices) → reality → social change (willingness of action).

However, the reality that the president attempts to create is that climate change is not real, and there are no climate change-associated threats. Therefore, this reality aims at making people to ignore the problem.

> Ideologies → (discursive practices) → reality → social change (willingness skipping the problem).

Each group aims to make the audience view their attitude as reality, which is the goal that each group try to achieve.

The governors and mayors representing one group in the context of this book try to make their attitude, that climate change is real and therefore the U.S. needs to remain in the Paris agreement, seen as reality (knowledge), and in turn they (the governors and mayors) try to make the other group's attitude (the president' attitude) viewed only as a (negative) ideology, and not reality. This leads us to investigate why the politicians aim at making their attitude to be viewed as a reality. In fact, the politician's attitude is an opinion, and this opinion remains a personal opinion or confined to a group until he/she discursively makes it seem a reality to the audience since when it is viewed as a reality it will has an effective impact on the addressee. In this respect, the politician can convince the addressee of his/her argument when he/she sounds right, so they can be foregrounded or backgrounded (pushed from attention of the recipient) according to Charteris-Black (2014, 2018), with the result that the audience can accept that opinion as a reality and the oppositional opinion as an ideology. As said earlier, according to many scholars (e.g., Lakoff & Johnson, 1980; Semino, 2005; Fairclough, 2014) the main goal of producing a text is persuasion and persuasion cannot be achieved unless the speaker/the writer sounds right (appeals to ethos) and sounding right requires the audience to view the speaker/writer's view as a reality and not only as an opinion for the purpose of triggering social change. As the figure indicates, *willingness of action* is the social change, which should be on the part of the government and people. Thus, it can be said that it (willingness of action) is represented by persuading all the states and governments that climate change is a real threat to the country and the world so the U.S. should remain in the Paris agreement and to respond to the problem instead of ignoring it. As for people' *willingness of action*, it is represented by calling the federal government for implementing more measures to tackle climate change starting from calling for keeping joining the Paris agreement.

This means that there is a social struggle between these groups, with each trying to make their attitude as a reality (so they legitimize their own attitude) and to make the other attitude as ideology (they delegitimize the other attitude). "Many environmental issues involve power struggles between opposing groups, and these struggles frequently take place in, and over, language." (Haig, 2001, p. 209). So, in this book, social struggle involves struggle over which attitude is to be viewed as a reality and which is to be viewed as an ideology. Thus, metaphors have an important role in de(legitimization). "Indeed, metaphor is often used both to legitimize and to delegitimize in the same text" (Charteris-Black, 2004, p. 17).

Because metaphor plays an important role in directing people's attention, many researchers (e.g., Mey, 2018) classified metaphors as positive or negative metaphors. With regards to the metaphors used by the governors, senators, and mayors, they are considered positive metaphors since these politicians used metaphors to stand with the environment through stressing to respond to climate change and protecting the environment from climate change and its consequences. The governors' and mayors' metaphors also raise people's awareness of the necessity of responding to the threats of climate change by implementing more measures to tackle this problem. Politicians employ many metaphors to reflect their strategies in a way that can be seen positively by the audience, and thereby underestimating the oppositional attitude (Charteris-Black, 2004; Fløttum, 2017).

Metaphors are positive or effective when they contribute to finding solutions to deal with climate change (such as War, Journey, Cleanliness and Construction). If I would have to choose a metaphor that politicians should more frequently use, it would be Journey metaphor because it is important to focus on how to deal with this problem on the long term due to the long nature of this problem. Thus, long-term measures are needed. It is not only important to start dealing with this problem. But there should be plans and policies that can be applied on the long term so we cope with this problem and its impacts. However, Construction metaphor is also important because this stresses cooperation. Such a task (dealing with climate change), is difficult, cannot be performed by a single state or country or even a few countries. This problem is global in nature. It is, therefore, important that politicians use metaphors that stress cooperation among governments. A mix of Journey and Construction metaphor is thus highly effective.

My implications are based on the following:

- Scientific evidence exists and says climate change is threatening to the environment.
- Climate change prompts implementing effective measures to cope with the problem, as a top priority.
- The environment including humans, plants and animals are all important.
- Texts produced by politicians on climate change either encourage people to address climate change or discourage people from addressing climate change.

Similarly, at the linguistic level, I also borrowed the words *foe* and *friend* terms from the researcher's works on metaphors (such as Mey, 2018) to classify metaphors used to express ideologies of the participants, analyzed in the study. "Metaphors are essential tools, needed for our survival in an environment that is not always user friendly" (Mey, 2018, p. 213). So, based on my ecosophy, the governors' and mayor's metaphorical expressions are positive (friends) metaphors, and, however, the other side uses negative (*foe)* metaphors. This classification is based on my perspective that the governors' and mayors' metaphors raise people's awareness of the necessity of responding to the threats of climate change by implementing more measures to tackle the problem. This reality is created through discourse of legitimization represented by metaphors placed in four (semantically related) groups from which conceptual metaphors are realized. Therefore, they are *ecologically constructive discourses*. Also, it is a call for the president and the federal government that they should deal with the problem to minimize its consequences, starting at keeping joining in the Paris agreement. Thus, these metaphors promote protect the environment and make the world safe. However, metaphors that call for the other attitude are viewed as *negative or foe metaphors* as they are used to express the belief that climate change is not real, and therefore they are *ecologically destructive discourses* (Table 9.2).

"In general, metaphoric expressions, whether positive or negative with respect to their ecological origin and force (think vs. danger) are ways of thinking and speaking that not only reflect, but also influence, our common social praxis" (Mey, 2018, p. 213).

Also, Semino (2008), who investigated discourse from a cognitive perspective, argues that "the relationship between discourses and ideologies is a dynamic one: discourses reflect particular ideologies, but also contribute to shape them and change them" (2008, p. 90). Thus, metaphor can mediate between discourse and ideologies. That is, according to Semino, discourse can construct conceptual metaphors which is in turn can produce or trigger ideologies. Leaders or politicians employ many metaphors to reflect their strategies in a way that can be seen positively by the audience and disparage or underestimate the oppositional attitude (Charteris-Black, 2004; Fløttum, 2017).

Thus, any efforts to treat climate change as those efforts made by such senators, governors and mayors in the study should be positively appraised.

Table 9.2 Discourses, ideologies and metaphors

Discourses	Ideologies	Metaphors (linguistic surface)
Ecologically constructive	Positive ideology	Friend (positive) metaphors
Ecologically destructive	Negative ideology	Foe (negative) metaphors

References

Carvalho, A. (2018). Climate Change and the Socio-Ecological Crisis. In A. F. Fill & H. Penz (Eds.), *The Routledge Handbook of Ecolinguistics* (pp. 487–499). Routledge.

Charteris-Black, J. (2004). *Politicians and Rhetoric. The Persuasive Power of Metaphor*. Palgrave Macmillan.

Charteris-Black, J. (2014). *Analyzing Political Speeches: Rhetoric, Discourse and Metaphor* (1st ed.). Palgrave Macmillan.

Charteris-Black, J. (2018). *Analyzing Political Speeches: Rhetoric, Discourse and Metaphor* (2nd ed.). Palgrave Macmillan.

Fairclough, N. (2014). *Language and Power* (3rd ed.). Routledge.

Fairclough, N., & Wodak, R. (1997). Critical Discourse Analysis. In T. VanDijk (Ed.), *Discourse as Social Interaction* (pp. 258–284). Sage.

Fløttum, K. (2017). Willingness of Action. In K. Fløttum (Ed.), *The Role of Language in the Climate Change Debate* (pp. 113–129). Routledge.

Gee, J. P. (2014). *An Introduction to Discourse Analysis: Theory and Method.* Routledge.

Haig, E. (2001). A Study of the Application of Critical Discourse Analysis to Ecolinguistics and the Teaching of Eco-literacy. *Critical Discourse Analysis for Ecolinguistics, 2*, 205–226.

Lakoff, G., & Johnson, M. (1980). *Metaphors We Live By*. The University of Chicago Press.

Mey, J. (2018). The Pragmatics of Metaphor: An Ecological View. In A. F. Fill & H. Penz (Eds.), *The Routledge Handbook of Ecolinguistics* (pp. 211–223). Routledge.

Naess, A. (1989). *Ecology, community, and lifestyle: Outline of an ecosophy.* Cambridge: Cambridge University Press.

Reisigl, M., & Wodak, R. (2008). The Discourse-Historical Approach (DHA). In R. Wodak & M. Meyer (Eds.), *Methods of Critical Discourse Analysis* (pp. 87–119). Sage.

Semino, E. (2005). The Metaphorical Construction of Complex Domains: The Case of Speech Activity in English. *Metaphor and Symbol, 1*, 35–70.

Semino, E. (2008). *Metaphor in Discourse*. Cambridge University Press.

Stibbe, A. (2015). *Ecolinguistics: Language, Ecology, and the Stories We Live By*. Routledge.

Stibbe, A. (2017). Critical Discourse Analysis and Ecology. In J. Flowerdew & J. E. Richardson (Eds.), *The Routledge Handbook of Critical Discourse Studies* (pp. 497–509). Routledge.

Stibbe, A. (2018). Positive Discourse Analysis: Rethinking Human Ecological Relationship. In A. F. Fill & H. Penz (Eds.), *The Routledge Handbook of Ecolinguistics* (pp. 165–178). Routledge.

Sum, N., & Jessop, B. (2013). *Towards a Cultural Political Economy: Putting Culture in Its Place in Political Economy*. Edward Elgar Publishing.

Van Dijk, T. (1997). *Discourse as Social Interaction*. Sage.

Van Dijk, T. (2015). Critical Discourse Analysis. In D. Tannen, H. E. Hamilton, & D. Shiffrin (Eds.), *The Handbook of Discourse Analysis* (pp. 466–485). Wiley Blackwell.

10

Contributions

The tenth chapter (contributions) summarizes the book's contribution to knowledge in three main areas: metaphor theory and analysis (developing conceptual metaphor theory) and contribution to climate change debate in the U.S. (e.g., framing climate change debate in the U.S.) as well as the book's contribution to Ecolinguistics. This chapter also summarizes the main findings and key points of this book so that the reader can have an overall view of how the metaphor analysis has been expanded and developed. This chapter serves as guidelines for the researchers pursuing metaphor studies as it includes directions for future research.

Contribution to Metaphor Theory and Analysis

This book developed metaphor theory and analysis to offer deeper understanding of how metaphor can trigger people through the metaphor opposition which is the second cognitive framework. I thus developed this theory to explain how metaphors can serve as a motivation that can push the addressee to take action on climate change or/and as a warning from its consequences, pushing the addressee to comply with the speaker/writer. However, whether motivation or warning, they can trigger the addressee to comply with the speaker's request which is implicated in

O. K. Al-Shboul, *The Politics of Climate Change Metaphors in the U.S. Discourse*,
https://doi.org/10.1007/978-3-031-19016-2_10

metaphor. That is, after the addressee's understanding of the topic through mapping, reasoning is an important process through which the addressee is warned about the consequences or/and is given a sense of encouragement. This depends on how metaphor is processed (which is embodied in two different image schemas: mapping and opposition) where he/she draws a consequence. Accordingly, concepts, making up each conceptual metaphor (e.g., Journey or War), which are classified as favored (positive) and unfavored (negative) are in a binary cognitive structure. For example, when climate change is conceptualized using metaphorical expressions of the Journey domain, the Journey metaphor can direct the addressee to make inferences that taking steps is necessary to moving forward not backward. If we do not move forward (the favored) by dealing with climate change, then we move backward (the unfavored). Also, if we do not move in the right direction (the favored), then we are in the wrong direction (the unfavored). These all possible inferences (forward or backward; the right path or the wrong path; destination or vagueness) are made by the addressee through metaphor opposition represented by this schema (B ↔ C). They thus direct him/her to make social change, which is the main goal of using metaphors by these politicians.

It can be argued that the concepts making up the (positive) favored concepts encourage the addressee to adopt what is called for by the speaker/writer, and the concepts making the unfavored (negative) concepts, serve as a warning, for the addressee of the consequences in case he/she did not comply with what the speaker/writer is calling for. Accordingly, positive (favored) and negative (unfavored) aspects both push the addressee to comply with the speaker/writer's argument. At a more advanced level of cognitive analysis, I combined the two theories (mapping and metaphor opposition) through 'feeding'. I named the new theory 'Metaphor Feeding theory'. The 'feeding' term is used in phonology to mean "to open the door for the other rule (e.g., sound change) to apply". Accordingly, in this book, I argue that mapping feeds metaphor opposition. That is to say, it opens the door for the metaphor opposition to apply (so that it is activated in the addressee's mind so that further inferences can be made).

To put it another way, the addressee's perception of metaphor involves two cognitive processes: understanding (the addressee perceives realities constructed by a metaphor) and reasoning (the addressee draws a consequence) with each process run by a different cognitive mechanism (mapping and opposition respectively). This leads the addressee to think of the problem (climate change or the Paris agreement) in certain ways. Accordingly, what triggers the hearer to act is the second part of the addressee's cognitive processing of metaphors which is the opposite sides (involving the opposing features), but understanding, which is via mapping, is the path to this trigger, so they are complementary in the sense that the first side A→ B helps the addressee understand the target domain, which is something necessary for triggering the addressee for further inferences (which is the second cognitive process-reasoning). The other image scheme (B ↔ C) leads the addressee to draw a consequence, which serves as an encouragement or warning that he/she perceives. Thus, it can be perceived as "you should … to avoid….".

Additionally, what makes the reader perceives warning or encouragement is the triggering of the opposing concepts. The main components of the mapping schema: is the source and target domain, while that of the opposition image schema is the target domain and the opposing concepts as shown in the Fig. 10.1.

Based on Metaphor Feeding theory: as shown in the figure above, this theory claims that while understanding is achieved via mapping, reasoning (drawing a conclusion) is achieved via opposition.

Conceptual structures/image schema	Resulting Cognitive process
Mapping	Understanding
Opposition	Reasoning

Fig. 10.1 Image schema and cognitive processes

Contribution to Studying Climate Change Debate in the U.S.

Despite the scientific evidence and consensus on climate change and its dangerous consequences, there has still been political disagreement over this environmental problem until this moment (Fløttum, 2017). That is, unlike in science, in politics climate change is debated as Man-made or is not; exists or does not, has disastrous consequences or has not and should be a top priority or should not. This is what this problem looks like in politics. Gunster (2012) maintains that climate change gets more complicated when it is debated in politics where politicians have more active access to public resources while, and subsequently, the voices of ordinary people become passive or less influential. "Therefore, climate politics almost always appears as part of the problem and rarely as part of the solution" (Gunster, 2012, p. 253).

This book advances research into climate change debate in the U.S. context. It offers insights into how the climate change debate is structured and can be studied. This book thus outlines the debate structure between politicians in the United States as two main opposing arguments, with regards to climate change: those who argue for dealing with climate change (they thus support the Paris climate agreement) and those who do not (they thus denounce the agreement). In this regard, this book therefore indicates the main focus of each side or group of those politicians: the core component of the President and politicians' argument (supporting the exit) is the Paris climate agreement. But the governors, senators' and mayors' argument (which opposes the exit) centers on the problem (climate change) itself rather than on the Paris agreement (and when they refer to this agreement, they refer to it as an important opportunity that the United States needs to keep). This can show that unlike the politicians opposing the exit, the President's (and all the politicians defending the exit) concern is about the deal, as being unfair and the United States suffers from this unfairness of this deal (they focus on the deal and its consequences for the U.S. rather than on climate change and its global consequences). Thus, the focuses of the two groups are substantially different. The politicians (the senators, governors and mayors) opposing the Paris climate exit call for the President and the federal government that they should deal with, and take, this problem as a top priority to minimize its consequences, starting at keeping joining in the Paris agreement, and consequently these metaphors used by these senators, governors and mayors promote protecting the environment.

In the light of metaphor, this book offers additional focuses of the debate of each side. In this regard, it investigates how climate change is conceptualized in the U.S. politics and in the media. One way to investigate this conceptualization is examining metaphor, employed by politicians. Dryzek (2013) says that metaphors are "rhetorical devices, deployed to convince listeners or readers by putting a situation in a particular light" (p. 19). Politicians draw on this feature for the sake of delivering their message to the audience. Charteris-Black (2005) found that all the politicians "make extensive use of metaphor" (p. 197). With regards to the metaphors used by the social actors, the President and the politicians (including the governors and senators) defending the exit from Paris agreement used more metaphors that are not used by the politicians (the senators, governors and mayors) opposing the exit, including Impediment, Unfairness, and Theft. This is because, unlike the senators, governors and mayors, the President and politicians who supported the exit focused more on the agreement itself than climate change and its consequences. They, therefore, used metaphors that suit this topic. They justified their decision to withdraw the U.S. from the Paris climate agreement by presenting climate change as non-threatening problem and it is not anthropogenic. They rather conceptualized the Paris agreement as being unfair to the country and as an impediment to the economic growth in the country. Besides, they used other metaphors drawn from the domain of War, Journey, Cleanliness, and Construction to defend their decision. The Paris agreement was frequently mentioned in their statements, they presented it as a real problem which needs to be renegotiated for the purpose of pursuing a better deal. However, the senators, governors and mayors opposed the withdrawal from the Paris climate agreement by conceptualizing and presenting climate change as a serious and threatening problem which needs to be urgently addressed (and Paris agreement is an effective way to tackle the problem), rather than arguing the President whether the agreement is fair or unfair. They focused on the impacts and consequences of climate change. Accordingly, this book can guide us to identify which politicians/groups make climate change a top priority. That is, politicians/groups that predominately use metaphors from certain source domains such as Journey, War and Construction on climate change are more concerned about climate change than politicians/groups that use metaphors from domains such as Unfairness and Business. This is because metaphors drawn from the source domains of War, Journey, Cleanliness and Construction emphasize dealing with climate change since they imply calling for action to address this problem. Metaphors drawn from Business, on the other hand, restrain action on climate change since they create excuses for delaying action or even not dealing with this environmental problem. The results indicate that climate change is mainly framed as a long-term and threatening problem that policy makers should seriously deal with. While for those politicians supporting the exit, the problem is framed along with the Paris agreement as unfair. However, it is not only what domain a metaphor is drawn from can determine if a politician/group (using that metaphor) supports or denounces responding to climate change (dealing with this problem as a top priority or not) but also this depends on what the purpose behind using that metaphor is. Accordingly, the focuses and domains from which the politicians draw metaphors are key factors in helping reveals their political and social goals (Table 10.1).

Table 10.1 Politicians' main focuses regarding climate change

Politicians	Main Focus
Politicians supporting the exit	Paris agreement
Politicians against exist	Climate change

Contribution to Ecolinguistics

This book contributes to ecolinguistics through my ecological philosophy (ecosophy) which I included in the ninth chapter. I set up a philosophical framework against which I identified metaphors as positive or negative. That is, the ones which promote policies on climate change are positive while those which do not are negative. In this book, Ecolinguistics thus tries to classify metaphors, as ideology loaded devices, to determine the politicians' stances towards the environmental issues investigated, thus I judged these (two) stances based on my assessment, which is in turn based on my beliefs about the issue investigated.

Accordingly, studying metaphor from ecolinguistic perspective can guide us to understand how politicians frame climate change, what social changes that they try to trigger and how they can affect the addressee's perceptions of this problem. Many experimental studies (e.g., Flusberg et al., 2017, 2018) on metaphor presented concrete evidence that metaphors can develop a sense of urgency in the participant. Thus, investigating this feature in particular using different approaches (qualitative and quantitative methods) can provide us with more comprehensive explanations of how climate change is conceptualized and what inferences and entailments that the addressee can make, and therefore how politicians frame this problem.

These politicians including the president, senators, governors and mayors used these metaphors in particular to create those realities discussed above and throughout this research about climate change and the Paris agreement. The results might have been different if I had been analyzing discourse produced by other groups or institutions discussing climate change. This of course depends on to a large extent on what agendas politicians hold and what realities they aim to create about climate change. That is, when a politician believes that climate change is anthropogenic, he/she would use metaphors that are different from those that are used by a politician who believes that climate change is nonanthropogenic. According to Fairclough (2014), different metaphors reflect different ideological loadings.

Summary and Further Research

In order to explain how research into metaphor and climate change may be advanced, I will summarize the key points in this book and then list important points for further research.

This book draws upon a mixed of methods to investigate politicians' reaction to the president's decision in politics (using qualitative analysis) and in the media (using corpus analysis) to explore how climate change is conceptualized in the politics of the United States from the perspective of Ecolinguistics and Critical Discourse Analysis. I analyzed all the statements and letters issued by the senators, governors and mayors who opposed the American president (Donald Trump) when he announced that he would withdraw the U.S. from the Paris climate agreement, which 195 countries around the world signed. The qualitative analysis employs metaphor analysis to investigate politicians' metaphorical choices made to influence public opinion and influence policies about climate change in the U.S.

I carried out an analysis at the cognitive level, using two cognitive frameworks: *the source-target mappings* and *metaphor opposition* to offer a more comprehensive explanations of how climate change is conceptualized and how inferences can be made and how metaphor can trigger people.

For the cognitive mapping, the features or concepts (which represent the physical or personal experience) associated with the source domains map (transferred) onto climate change (in terms of what and how climate change is and should be dealt with) as explained in this chapter. Consequently, these source domains foreground some aspects of realities about the less familiar domain (climate change) in ways that make the target domain conceptualized by these source domains: climate change in terms of journey; climate change in terms of war; climate change in terms of cleanliness; climate change in terms of construction. Conceptual metaphors made by such source domains shape then reasoning about the problem and make it accessible to the reader, according to many researchers (e.g., Charteris-Black, 2014, 2018; Flusberg et al., 2018; Hauser &

Schwarz, 2014; Larson, 2011; Mio, 1997; Sweetser, 2017; Thibodeau et al., 2017).

But why mapping occurs (how is it triggered)? Mapping is triggered because of semantic clash, that the addressee perceives, between the source and target domain on the semantic level (Charteris-Black, 2014, 2018; Kövecses, 2016; Musloff, 2016), with the result that some notions of the source domain map into the target domain. With regards to the data of this book, the governors and mayors used many metaphorical expressions that semantically or literally appear to be unfit to be used to describe the abstract domain (climate change) since these expressions are drawn from concrete (everyday experiences) domains while climate change is, on contrary, abstract and quite different. Journey, War, Cleanliness and Construction seem to have no any semantic (literal) ties with this domain. For example, when the politician says *we need to fight climate change*. Here, the lexical verb *fight* is semantically used with living beings that are regarded as enemies, but not with inanimate objects. This means that there is a clash on the semantic or literal level. Consequently, in order for the addressee to process the meaning in this expression involving such a clash, some conceptual notions or elements associated with the verb *fight* map onto the climate change domain, and thereby climate change is perceived as an enemy that needs to be fought. Also, because it is an enemy, the speaker emphasizes that governments/policy makers should respond to climate change urgently to avoid suffering from many serious consequences. Thus, what leads to such inferential mappings is the semantic clash (the literal meaning of war being inconsistent with climate change) that the addressee perceives on the literal level.

Additionally, the semantic clash between the source and the target domain is functional since this semantic clash points out to the long-term nature of the phenomenon with regards to Journey metaphors (since journey focuses on movement to destination); it points out to the seriousness of the problem with regards to War metaphors (since war causes fear and anxiety); it shows that the problem is Man-made due to emissions with regards to Cleanliness metaphors (since cleanliness emphasizes adopting a new type of technology that does not cause pollution); and it points out to the difficulty of the problem with regards to

Construction metaphors (since construction requires cooperation and contribution from workers with different specialties).

The President and politicians supporting the exit used seven types of metaphors, including Unfairness, Impediment, War, Journey, Cleanliness, Construction and Theft. The Most common ones are Impediment and Unfairness. Both of which belong to the larger domain, which is Business or Economy. By way of explanation, many metaphorical expressions belonging to these domains (Unfairness and Impediment) are used to shape the reality that the Paris agreement imposes unequal obligations to its members, undermines the economy of the country and blocks the developments. The other metaphors that they (the President and politicians supporting the exit) used, which are War, Journey, Cleanliness and Construction are not aimed at seeking out solutions to tackle climate change. Rather, he drew on them to defend his decision and to demonstrate that he works hard and he is willing to cooperate with other members/ parties to sign a new deal with new terms. For the governors and mayors, on the other hand, they used four types of metaphors: Journey, War, Cleanliness and Construction. The most common metaphor used in this book is Journey and the least common was Construction. Although the President and governors and mayors used metaphors of Journey, War, Cleanliness and Construction, they were not used with the same aim (as when they were used by the president). That is, by using these metaphors, the governors and mayors responded to the President by stressing that climate change is a top priority and the government needs to implement more effective measures and procedure to deal with this problem and its consequences. They assert that the Paris climate agreement is one effective way to address this problem and its consequences. Simply put, the President and other politicians supporting the Paris agreement exit believe that the Paris agreement is a problem (a problem in and of itself). Withdrawing the U.S. from the Paris climate agreement is thus a must. They predominantly drew on many metaphors to construct this reality. The governors and mayors, on the other hand, predominantly framed climate change as a long-term problem and as a security threat via drawing overwhelmingly on metaphors of Journey and War. They believe that the Paris climate agreement is part of the solution. Withdrawing the U.S. from the Paris climate agreement, therefore, is a big mistake.

Framing this issue as a threat is what the media and politicians focused on. That is, the media most frequently used terms from the domain of war and also the governors and mayors very frequently drew on this domain. Simply put, the war terms were metaphorically used in the media just as the politicians used these terms to talk about climate change. Moreover, the corpus analysis reveals that the world and U.S. media focused on climate change in terms of its consequences, as demonstrated by the most frequent nouns associated with climate change: *effects and impact(s)*, and accordingly it highlights actions of urgency, needed to address climate change and its consequences, lexicalized by these verbs (metaphorical expressions of war) *combat, fight and tackle*. These verbs substantialize actions of urgency just as metaphors of War used by the senators, governors and mayors to conceptualize responding to climate change. Additionally, the statistical results suggest that the decision sparked heated debate in the U.S. and world media. That is, the differences of the frequency of occurrences of the climate change/ global warming and Paris agreement, Paris accord and Paris deal terms before and after the announcement are statistically significant in the U.S. and world media (as indicated by the Log-likelihood results). The gaps, however, are bigger and more meaningful in the U.S media (as indicated by the Log Ratio results). I investigated if the world and U.S. media devoted a large amount of attention to this announcement through looking at the frequency of occurrences of specific terms including *climate change/global warming, Paris agreement/accord/deal*. The statistical findings reveal that the media paid special attention to this event, especially the U.S. media, as evidenced through the significant and meaningful differences in the frequency of occurrences of those items before and after the announcement. Thus, this reveals that the U.S. media devoted a larger amount of attention to this event (the announcement of the U.S. president to withdraw the U.S. from the Paris agreement).

One important aspect with regards to metaphors investigated in this book is urgency. I found a scale for that purpose: the conceptual metaphors investigated in this book is scaled in terms of urgency starting from War, Cleanliness, Construction to Journey. An important finding is that the governors in the West stressed stronger urgency than the other regions. Additionally, it was concluded that the region has an important role to play in pushing politicians to oppose against the decision. That is, many more politicians from the Northeast and West regions than the other remaining regions responded to the decision of withdrawing the U.S. from the Paris climate agreement. This can be attributed to the fact that these regions are the most affected areas because of climate change (as reported in the national climate assessment report, 2018).

This book also used corpus analysis to investigate the changes in the discourse about climate change in the media before and after Donald Trump's announcement of the U.S. withdrawal from the Paris agreement in 2017. According to the findings of the corpus-based analysis, climate change is mainly framed as a threatening problem. This is evidenced through the most collocated terms, with the climate change term, which belong to the source domain of War (e.g., fight and combat). This reveals that the media, to a large extent, deals with the problem in much the same way as the governors and mayors, who very widely used metaphorical expressions drawn from the source domain of war. However, the governors and mayors drew more frequently on the source domain of Journey than the source domain of war. With the predominant use of metaphors of Journey by the governors and mayors, they focused on moving toward long-term solutions (acting), rather than just identifying the nature of this problem (as being threatening, and hence promptly addressing the problem). By way of explanation, the governors and mayors drew most frequently on the source domain of Journey, which emphasizes how to solve the problem on the long term. This includes what measures and procedures (and how they) should be employed on the long term rather than the medium and short term. Journey focuses, as argued by many researchers (e.g., Atanasova & Koteyko, 2017; Semino, 2008), on moving toward destination, rather than destination itself, and subsequently on implementing effective measures and procedures on the long term to deal with climate change.

Based on the issues investigated in this book (including metaphor theory and climate change debate), the area of this research may be advanced by:

1. Conducting a contrastive study to analyze the data from politicians at two or more different countries to find out how language is employed

by those groups and what language features (e.g., hyperbole) and patterns (e.g., nomination) can be recognized, that can play an important role in creating realities. This can contribute to understanding how this problem (climate change) is discursively constructed at a larger political structure.

2. Conducting a study on the same sample or a larger sample to investigate voices through employing polyphony theory from CDA perspective to reveal what are implicit or implied voices in the politicians' language. Politicians have access to many resources (e.g., debates and press conferences) that ordinary people do not. They, therefore, speak on behalf of others as well. It is worth investigating how language is used by these politicians to construct voices.
3. Conducting a sociolinguistic study to investigate if sociolinguistic variables affect speakers'/writers' linguistic choices when discussing environmental problems. For example, if gender influences the politician's metaphorical choices on climate change. This is one way to examine how metaphors are employed by both genders in sociopolitical contexts. That is, how male and female politicians use language on climate change and what metaphorical patterns that can be identified. From what source domains does each gender more frequently uses metaphors (e.g., Journey, War or Construction)?.
4. Using corpus analysis to conduct a comparative study between two different groups: political discourse and business discourse to investigate how climate change is represented by these groups (of these different backgrounds). For example, what discourses do each group focus on? Do politicians frame climate change as an economic or as threatening problem? Do companies frame climate change as an economic or threatening problem?.
5. Conducting an experimental study to investigate if there are specific metaphors that can develop a stronger sense of urgency in the participant to address climate change than others. This requires surveying a big number of people who need to read texts describing climate change through metaphors, with each text describing climate change with a specific type of metaphors (e.g., War or Journey) to investigate how participants perceive this phenomenon when constructed and presented differently using different metaphors. A scaled questionnaire is

then conducted to ask these participants in each group about their perception (for example, if it would be necessary to respond to this problem promptly). This experimental study is to offer empirical evidence if War metaphor can develop within the reader a sense of urgency in comparison with other types of metaphors such as Journey and Construction.

References

Atanasova, D., & Koteyko, N. (2017). Metaphors in online editorials and op-ed in climate change, 2006–2013: A study of Germany, the United Kingdom, the United States. In K. Flottum (Eds.), *The Role of Language in the Climate Change* (pp. 71–89). New York, NY: Routledge.

Charteris-Black, J. (2005). *Politicians and Rhetoric. The Persuasive Power of Metaphor*. Palgrave Macmillan.

Charteris-Black, J. (2014). *Analyzing Political Speeches: Rhetoric, Discourse and Metaphor* (1st ed.). Palgrave Macmillan.

Charteris-Black, J. (2018). *Analyzing Political Speeches: Rhetoric, Discourse and Metaphor* (2nd ed.). Palgrave Macmillan.

Dryzek, J. (2013). *The Politics of the Earth: Environmental Discourses* (3rd ed.). Oxford University Press.

Fairclough, N. (2014). *Language and power* (3rd ed.). Routledge.

Fløttum, K. (2017). Willingness of Action. In K. Fløttum (Ed.), *The Role of Language in the Climate Change Debate* (pp. 113–129). Routledge.

Flusberg, S., Matlock, T., & Thibodeau, P. (2017). Metaphors for the War (or Race) Against Climate Change. *Environmental Communication, 6*, 769–783.

Flusberg, S., Matlock, T., & Thibodeau, P. (2018). War Metaphors in Public Discourse. *Metaphor and Symbol, 33*(1), 1–18.

Gunster, S. (2012). Visions of Climate Politics in Alternative Media. In A. Carvalho & T. R. Peterson (Eds.), *Climate Change Politics: Communication and Public Engagement* (pp. 247–277). Cambria Press.

Hauser, D. J., & Schwarz, N. (2014). The War on Prevention Bellicose Cancer Metaphors Hurt (Some) Prevention Intentions. *Personality and Social Psychology Bulletin, 41*(1), 66–77.

Kövecses, Z. (2016). A View of Mixed Metaphor Within a Conceptual Metaphor Theory Framework. In R. W. Gibbs (Ed.), *Mixing Metaphor* (pp. 3–15). John Benjamins Publishing Company.

Larson, B. (2011). *Metaphors for Environmental Sustainability: Redefining Our Relationship with Nature*. Yale University Press.

Mio, J. S. (1997). Metaphor and Politics. *Metaphor and Symbol, 12*(2), 113–133. https://doi.org/10.1207/s15327868ms1202_2

Musloff, A. (2016). *Political Metaphor Analysis: Discourse and Scenarios*. Bloomsbury Publishing.

Semino, E. (2008). *Metaphor in discourse*. Cambridge: Cambridge University Press.

Sweetser, E. (2017). Conceptual Mappings. In B. Dancygier (Ed.), *The Cambridge Handbook of Cognitive Linguistics* (pp. 379–384). Cambridge University Press.

Thibodeau, P. H., Hendricks, R. K., & Boroditsky, L. (2017). How Linguistic Metaphor Scaffolds Reasoning. *Trends in Cognitive Sciences, 21*(11), 852–863.

United Nations. The Paris agreement. https://unfccc.int/files/essential_background/convention/application/pdf/english_paris_agreement.pdf

United States Climate Alliance. https://www.usclimatealliance.org/alliance-principles

USA Gov. https://www.usa.gov/

Appendix A: The Paris Agreement

Paris Agreement

12 December 2015

ENTRY INTO FORCE: 4 November 2016, in accordance with article 21(1). The Agreement enters into force on the thirtieth day after the date on which at least 55 Parties to the Convention accounting in total for at least an estimated 55 per cent of the total global greenhouse gas emissions have deposited their instruments of ratification, acceptance, approval or accession. REGISTRATION: 4 November 2016, No. 54113.

STATUS: Signatories: 195. Parties: 189.

TEXT: C.N.63.2016.TREATIES-XXVII.7.d of 16 February 2016 (Opening for signature) and C.N.92.2016.TREATIES-XXVII.7.d of 17 March 2016 (Issuance of Certified True Copies).

Note: The Paris Agreement was adopted on 12 December 2015 at the twenty-first session of the Conference of the Parties to the United Nations Framework Convention on Climate Change held in Paris from 30 November to 13 December 2015. In accordance with its article 20, the Agreement shall be open for signature at the United Nations Headquarters in New York from 22 April 2016 until 21 April 2017 by States and

O. K. Al-Shboul, *The Politics of Climate Change Metaphors in the U.S. Discourse*,
https://doi.org/10.1007/978-3-031-19016-2

regional economic integration organizations that are Parties to the United Nations Framework Convention on Climate Change.

Participant Signature Ratification, Acceptance(A), Approval (AA), Accession(a)

Afghanistan.. 22 Apr 2016 15 Feb 2017
Albania.. 22 Apr 2016 21 Sep 2016
Algeria.. 22 Apr 2016 20 Oct 2016
Andorra... 22 Apr 2016 24 Mar 2017
Angola.. 22 Apr 2016
Antigua and Barbuda................................. 22 Apr 2016 21 Sep 2016
Argentina.. 22 Apr 2016 21 Sep 2016
Armenia.. 20 Sep 2016 23 Mar 2017
Australia.. 22 Apr 2016 9 Nov 2016
Austria.. 22 Apr 2016 5 Oct 2016
Azerbaijan.. 22 Apr 2016 9 Jan 2017
Bahamas... 22 Apr 2016 22 Aug 2016
Bahrain... 22 Apr 2016 23 Dec 2016
Bangladesh... 22 Apr 2016 21 Sep 2016
Barbados.. 22 Apr 2016 22 Apr 2016
Belarus... 22 Apr 2016 21 Sep 2016
A Belgium..22 Apr 2016 6 Apr 2017
Belize... 22 Apr 2016 22 Apr 2016
Benin..22 Apr 2016 31 Oct 2016
Bhutan..22 Apr 2016 19 Sep 2017
Bolivia (Plurinational State of)....................22 Apr 2016 5 Oct 2016
Bosnia and Herzegovina.............................22 Apr 2016 16 Mar 2017
Botswana... 22 Apr 2016 11 Nov 2016

Participant Signature Ratification, Acceptance(A), Approval (AA), Accession(a)

Brazil.. 22 Apr 2016 21 Sep 2016
Brunei Darussalam...................................... 22 Apr 2016 21 Sep 2016
Bulgaria.. 22 Apr 2016 29 Nov 2016
Burkina Faso...22 Apr 2016 11 Nov 2016
Burundi...22 Apr 2016 17 Jan 2018
Cabo Verde...22 Apr 2016 21 Sep 2017
Cambodia..22 Apr 2016 6 Feb 2017

Cameroon..22 Apr 2016 29 Jul 2016
Canada.. 22 Apr 2016 5 Oct 2016
Central African Republic........................... 22 Apr 2016 11 Oct 2016
Chad.. 22 Apr 2016 12 Jan 2017
Chile.. 20 Sep 2016 10 Feb 2017
China...22 Apr 2016 3 Sep 2016
Colombia.. 22 Apr 2016 12 Jul 2018
Comoros.. 22 Apr 2016 23 Nov 2016
Congo.. 22 Apr 2016 21 Apr 2017
Cook Islands... 24 Jun 2016 1 Sep 2016
Costa Rica.. 22 Apr 2016 13 Oct 2016
Côte d'Ivoire.. 22 Apr 2016 25 Oct 2016
Croatia... 22 Apr 2016 24 May 2017
Cuba..22 Apr 2016 28 Dec 2016
Cyprus..22 Apr 2016 4 Jan 2017
Czech Republic... 22 Apr 2016 5 Oct 2017
XXVII 7 D. ENVIRONMENT 2
Participant Signature Ratification, Acceptance(A), Approval(AA), Accession(a)
Democratic People's Republic of Korea..........22 Apr 2016 1 Aug 2016
Democratic Republic of the Congo...............22 Apr 2016 13 Dec 2017
Denmark1..22 Apr 2016 1 Nov 2016
AA Djibouti...22 Apr 2016 11 Nov 2016
Dominica.. 22 Apr 2016 21 Sep 2016
Dominican Republic.................................. 22 Apr 2016 21 Sep 2017
Ecuador.. 26 Jul 2016 20 Sep 2017
Egypt... 22 Apr 2016 29 Jun 2017
El Salvador... 22 Apr 2016 27 Mar 2017
Equatorial Guinea...................................... 22 Apr 2016 30 Oct 2018
Eritrea.. 22 Apr 2016
Estonia... 22 Apr 2016 4 Nov 2016
Eswatini... 22 Apr 2016 21 Sep 2016
Ethiopia.. 22 Apr 2016 9 Mar 2017
European Union.. 22 Apr 2016 5 Oct 2016
Fiji .... 22 Apr 2016 22 Apr 2016
Finland... 22 Apr 2016 14 Nov 2016

France.. 22 Apr 2016 5 Oct 2016
Gabon..22 Apr 2016 2 Nov 2016
Gambia..26 Apr 2016 7 Nov 2016
Georgia.. 22 Apr 2016 8 May 2017
AA Germany...22 Apr 2016 5 Oct 2016
Ghana..22 Apr 2016 21 Sep 2016
Greece.. 22 Apr 2016 14 Oct 2016
Grenada..22 Apr 2016 22 Apr 2016
Guatemala..22 Apr 2016 25 Jan 2017
Guinea..22 Apr 2016 21 Sep 2016
Guinea-Bissau... 22 Apr 2016 22 Oct 2018
Guyana..22 Apr 2016 20 May 2016
Haiti..22 Apr 2016 31 Jul 2017
Honduras..22 Apr 2016 21 Sep 2016
Hungary... 22 Apr 2016 5 Oct 2016
Iceland... 22 Apr 2016 21 Sep 2016
A India..22 Apr 2016 2 Oct 2016
Indonesia..22 Apr 2016 31 Oct 2016
Iran (Islamic Republic of)........................... 22 Apr 2016
Iraq...8 Dec 2016
Ireland... 22 Apr 2016 4 Nov 2016
Israel... 22 Apr 2016 22 Nov 2016
Italy... 22 Apr 2016 11 Nov 2016
Jamaica.. 22 Apr 2016 10 Apr 2017

Participant Signature Ratification, Acceptance(A), Approval (AA), Accession(a)

Japan... 22 Apr 2016 8 Nov 2016 A
Jordan..22 Apr 2016 4 Nov 2016
Kazakhstan...2 Aug 2016 6 Dec 2016
Kenya...22 Apr 2016 28 Dec 2016
Kiribati...22 Apr 2016 21 Sep 2016
Kuwait...22 Apr 2016 23 Apr 2018
Kyrgyzstan... 21 Sep 2016 18 Feb 2020
Lao People's Democratic Republic...............22 Apr 2016 7 Sep 2016
Latvia...22 Apr 2016 16 Mar 2017
Lebanon...22 Apr 2016 5 Feb 2020

Lesotho.. 22 Apr 2016 20 Jan 2017
Liberia..22 Apr 2016 27 Aug 2018
Libya... 22 Apr 2016
Liechtenstein.. 22 Apr 2016 20 Sep 2017
Lithuania... 22 Apr 2016 2 Feb 2017
Luxembourg.. 22 Apr 2016 4 Nov 2016
Madagascar.. 22 Apr 2016 21 Sep 2016
Malawi... 20 Sep 2016 29 Jun 2017
Malaysia.. 22 Apr 2016 16 Nov 2016
Maldives.. 22 Apr 2016 22 Apr 2016
Mali...22 Apr 2016 23 Sep 2016
Malta...22 Apr 2016 5 Oct 2016
Marshall Islands...22 Apr 2016 22 Apr 2016
Mauritania..22 Apr 2016 27 Feb 2017
Mauritius..22 Apr 2016 22 Apr 2016
Mexico... 22 Apr 2016 21 Sep 2016
Micronesia (Federated States of)................22 Apr 2016 15 Sep 2016
Monaco..22 Apr 2016 24 Oct 2016
Mongolia..22 Apr 2016 21 Sep 2016
Montenegro..22 Apr 2016 20 Dec 2017
Morocco...22 Apr 2016 21 Sep 2016
Mozambique..22 Apr 2016 4 Jun 2018
Myanmar..22 Apr 2016 19 Sep 2017
Namibia...22 Apr 2016 21 Sep 2016
Nauru..22 Apr 2016 22 Apr 2016
Nepal...22 Apr 2016 5 Oct 2016
Netherlands2..22 Apr 2016 28 Jul 2017 A
New Zealand3 ....22 Apr 2016 4 Oct 2016
Nicaragua...23 Oct 2017 a
Niger... 22 Apr 2016 21 Sep 2016
Nigeria... 22 Sep 2016 16 May 2017
XXVII 7 D. ENVIRONMENT 3

Participant Signature Ratification, Acceptance(A), Approval(AA), Accession(a)

Niue.. 28 Oct 2016 28 Oct 2016
North Macedonia...22 Apr 2016 9 Jan 2018

Norway.. 22 Apr 2016 20 Jun 2016
Oman...22 Apr 2016 22 May 2019
Pakistan...22 Apr 2016 10 Nov 2016
Palau..22 Apr 2016 22 Apr 2016
Panama...22 Apr 2016 21 Sep 2016
Papua New Guinea.....................................22 Apr 2016 21 Sep 2016
Paraguay... 22 Apr 2016 14 Oct 2016
Peru..22 Apr 2016 25 Jul 2016
Philippines..22 Apr 2016 23 Mar 2017
Poland..22 Apr 2016 7 Oct 2016
Portugal..22 Apr 2016 5 Oct 2016
Qatar...22 Apr 2016 23 Jun 2017
Republic of Korea..22 Apr 2016 3 Nov 2016
Republic of Moldova...................................21 Sep 2016 20 Jun 2017
Romania..22 Apr 2016 1 Jun 2017
Russian Federation...................................... 22 Apr 2016 7 Oct 2019 A
Rwanda...22 Apr 2016 6 Oct 2016
Samoa...22 Apr 2016 22 Apr 2016
San Marino..22 Apr 2016 26 Sep 2018
Sao Tome and Principe................................22 Apr 2016 2 Nov 2016
Saudi Arabia..3 Nov 2016 3 Nov 2016
Senegal...22 Apr 2016 21 Sep 2016
Serbia..22 Apr 2016 25 Jul 2017
Seychelles.. 25 Apr 2016 29 Apr 2016
Sierra Leone...22 Sep 2016 1 Nov 2016
Singapore..22 Apr 2016 21 Sep 2016
Slovakia..22 Apr 2016 5 Oct 2016
Slovenia.. 22 Apr 2016 16 Dec 2016
Solomon Islands..22 Apr 2016 21 Sep 2016
Somalia...22 Apr 2016 22 Apr 2016
South Africa...22 Apr 2016 1 Nov 2016
South Sudan..22 Apr 2016
Spain...22 Apr 2016 12 Jan 2017
Sri Lanka...22 Apr 2016 21 Sep 2016
St. Kitts and Nevis..22 Apr 2016 22 Apr 2016

Participant Signature Ratification, Acceptance(A), Approval (AA), Accession(a)

St. Lucia..22 Apr 2016 22 Apr 2016
St. Vincent and the Grenadines....................22 Apr 2016 29 Jun 2016
State of Palestine...................................22 Apr 2016 22 Apr 2016
Sudan..22 Apr 2016 2 Aug 2017
Suriname...22 Apr 2016 13 Feb 2019
Sweden...22 Apr 2016 13 Oct 2016
Switzerland..22 Apr 2016 6 Oct 2017
Syrian Arab Republic.................................13 Nov 2017 a
Tajikistan...22 Apr 2016 22 Mar 2017
Thailand...22 Apr 2016 21 Sep 2016
Timor-Leste..22 Apr 2016 16 Aug 2017
Togo...19 Sep 2016 28 Jun 2017
Tonga..22 Apr 2016 21 Sep 2016
Trinidad and Tobago..................................22 Apr 2016 22 Feb 2018
Tunisia..22 Apr 2016 10 Feb 2017
Turkey...22 Apr 2016
Turkmenistan...23 Sep 2016 20 Oct 2016
Tuvalu...22 Apr 2016 22 Apr 2016
Uganda...22 Apr 2016 21 Sep 2016
Ukraine..22 Apr 2016 19 Sep 2016
United Arab Emirates................................ 22 Apr 2016 21 Sep 2016 A
United Kingdom
of Great Britain
and Northern Ireland.................................22 Apr 2016 18 Nov 2016
United Republic of Tanzania..........................22 Apr 2016 18 May 2018
United States of America4............................22 Apr 2016 3 Sep 2016 A
Uruguay..22 Apr 2016 19 Oct 2016
Uzbekistan...19 Apr 2017 9 Nov 2018
Vanuatu..22 Apr 2016 21 Sep 2016
Venezuela (Bolivarian Republic of)........... ..22 Apr 2016 21 Jul 2017
Viet Nam...22 Apr 2016 3 Nov 2016
AA Yemen...23 Sep 2016
Zambia...20 Sep 2016 9 Dec 2016
Zimbabwe...22 Apr 2016 7 Aug 2017

Declarations (Unless otherwise indicated, the declarations were made upon ratification, acceptance, approval or accession.)

BELGIUM

"This signature engages also the Walloon Region, the Flemish Region and the Brussels-Capital Region."

BULGARIA

"The Republic of Bulgaria recognizes that in accordance with Article 9, paragraph 1, of the Paris

Agreement developed country Parties shall provide financial resources to assist developing country Parties with respect to both mitigation and adaptation in continuation of their existing obligations under the Convention. In this context the Republic of Bulgaria notes that as a Party to the United Nations Framework Convention on Climate Change Bulgaria is not included in Annex II."

CHINA

"In accordance with the Basic Law of the Hong Kong Special Administrative Region of the People's Republic of China and the Basic Law of the Macao Special Administrative Region of the People's Republic of China, the Government of the People's Republic of China decides that the Agreement applies to the Hong Kong Special Administrative Region and the Macao Special Administrative Region of the People's Republic of China."

COOK ISLANDS

"The Government of the Cook Islands declares its understanding that acceptance of the Paris Agreement and its application shall in no way constitute a renunciation of any rights under international law concerning State responsibility for the adverse effects of climate change and that no provision in the Paris Agreement can be interpreted as derogating from principles of general international law or any claims or rights concerning compensation due to the impacts of climate change. The Government of the Cook Islands further declares that, in light of the best available scientific information and assessment on climate change and its impacts, it considers the emissions reduction obligations in the aforesaid Paris Agreement to be inadequate to prevent a global temperature stabilisation level at or above 1.5 degrees Celsius relative to pre-industrial levels and as a consequence, such emissions will have severe implications for our national interests."

EUROPEAN UNION

"Declaration by the Union made in accordance with Article 20(3) of the Paris Agreement The following States are at present Members of the European Union: the Kingdom of Belgium, the Republic of Bulgaria, the Czech Republic, the Kingdom of Denmark, the Federal Republic of Germany, the Republic of Estonia, Ireland, the Hellenic Republic, the Kingdom of Spain, the French Republic, the Republic of Croatia, the Italian Republic, the Republic of Cyprus, the Republic of Latvia, the Republic of Lithuania, the Grand Duchy of Luxembourg, Hungary, the Republic of Malta, the Kingdom of the Netherlands, the Republic of Austria, the Republic of Poland, the Portuguese Republic, Romania, the Republic of Slovenia, the Slovak Republic, the Republic of Finland, the Kingdom of Sweden, the United Kingdom of Great Britain and Northern Ireland. The European Union declares that, in accordance with the Treaty on the Functioning of the European Union, and in particular Article 191 and Article 192(1) thereof, it is competent to enter into international agreements, and to implement the obligations resulting therefrom, which contribute to the pursuit of the following objectives:- preserving, protecting and improving the quality of the environment;—protecting human health;—prudent and rational utilisation of natural resources;—promoting measures at international level to deal with regional or worldwide environmental problems, and in particular combating climate change.

... The European Union will continue to provide information, on a regular basis on any substantial modifications in the extent of its competence, in accordance with Article 20(3) of the Agreement."

INDIA

"The Government of India declares its understanding that, as per its national laws; keeping in view its development agenda, particularly the eradication of poverty and provision of basic needs for all its citizens, coupled with its commitment to following the low carbon path to progress, and on the assumption of unencumbered availability of cleaner sources of energy and technologies and financial resources from around the world; and based on a fair and ambitious assessment of global commitment to combating climate change, it is ratifying the Paris Agreement."

MARSHALL ISLANDS

"…the Government of the Republic of the Marshall Islands declares its understanding that ratification of the Paris Agreement shall in no way constitute a renunciation of any rights under any other laws, including international law, and the communication depositing the Republic's instrument of ratification shall include a declaration to this effect for international record; FURTHERMORE, the Government of the Republic of the Marshall Islands declares that, in light of best scientific information and assessment on climate change and its impacts, it considers the emission reduction obligations in Article 3 of the Kyoto Protocol, the Doha Amendment and the aforesaid Paris Agreement to be inadequate to prevent global temperature increase of 1.5 degrees Celsius above pre-Industrial levels and as a consequence, will have severe implications for our national interests…"

MEXICO

"in accordance with their national legal framework, and in consideration of the best and most up-to-date scientific information available and incorporated by the Intergovernmental Panel on Climate Change, the United Mexican States understands greenhouse gas emissions to mean the release into the atmosphere of greenhouse gases and/or their precursors and aerosols into the atmosphere, including, where applicable, greenhouse compounds, within a specific area and during a specific period of time."

MICRONESIA (FEDERATED STATES OF)

"The Government of the Federated States of Micronesia declares its understanding that its ratification of the Paris Agreement does not constitute a renunciation of any rights of the Government of the Federated States of Micronesia under international law concerning State responsibility for the adverse effects of climate change, and that no provision in the Paris Agreement can be interpreted as derogating from principles of general international law or any claims or rights concerning compensation and liability due to the adverse effects of climate change; and The Government of the Federated States of Micronesia further declares that, in light of the best available scientific information and assessments on climate change and its impacts, it considers the emission reduction obligations in the Paris Agreement to be inadequate to prevent a global temperature increase above 1.5 degrees Celsius relative to pre-industrial levels, and as a consequence, such emissions will have severe implications for the national interests of the Government of the Federated States of Micronesia."

NAURU

"… the Government, of Nauru declares its understanding that the ratification of the Agreement shall in no way constitute a renunciation of any rights under international law concerning State responsibility [for] the adverse effects of climate change. FURTHER, the Government of Nauru declares that no provisions in the Agreement can be interpreted as derogating from the principles of general international law. AND FURTHER, the Government of Nauru declares its understanding that Article 8 and decision 1/CP.21, paragraph 51 in no way limits the ability of Parties to UNFCCC or the Agreement to raise, discuss, or address any present or future concerns regarding the issues of liability and compensation. The Republic of Nauru put forth its concern intended to recognize and acknowledge its national interest…"

NETHERLANDS

"The Kingdom of the Netherlands, for the European part of the Netherlands, declares in accordance with Article 14, paragraph 2, of the United Nations Framework Convention on Climate Change in conjunction with Article 24 of the Paris Agreement, that it accepts both means of dispute settlement referred to in that paragraph as compulsory in relation to any Party accepting one or both means of dispute settlement."

NIUE

"The Government of Niue declares its understanding that acceptance of the Paris Agreement and its application shall in no way constitute a renunciation of any rights under international law concerning State responsibility for the adverse effects of climate change and that no provision in the Paris Agreement can be interpreted as derogating from principles of general international law or any claims or rights concerning compensation due to the impacts of climate change. The Government of Niue further declares that, in light of the best available scientific information and assessment on climate change and its impacts, it considers the emissions reduction obligations in the aforesaid Paris Agreement to be inadequate to prevent a global temperature stabilisation level at or above

1.5 degrees Celsius relative to pre-industrial levels and as a consequence, such emissions will have severe implications for our national interests."

PHILIPPINES

"THAT it is the understanding of the Government of the Republic of the Philippines that its accession to and the implementation of the Paris Agreement shall in no way constitute a renunciation of rights under any local and international laws or treaties, including those concerning State responsibility for loss and damage associated with the adverse effects of climate change; THAT, the accession to and implementation of the Paris Agreement by the Republic of the Philippines is for the purpose of supporting the country's national development objectives and priorities such as sustainable industrial development, the eradication of poverty and provision of basic needs, and securing social and climate justice and energy security for all its citizens."

POLAND

"The Government of the Republic of Poland recognizes that under Article 9 paragraph 1 of the Paris Agreement developed country Parties shall provide financial resources to assist developing country Parties with respect to both mitigation and adaptation in continuation of their existing obligations under the Convention. In this context the Government of the

Republic of Poland notes that Poland is a Party to the United Nations Framework Convention on Climate Change not included in Annex II."

RUSSIAN FEDERATION

1. The Russian Federation recognizes that, in accordance with paragraph 1 of Article 9 of the Agreement, developed country Parties shall provide financial resources to assist developing country Parties with respect to both mitigation of climate change and adaptation to it in continuation of their existing obligations under the United Nations Framework Convention on Climate Change of 9 May 1992 (hereinafter referred to as "the Convention"). In this context, the Russian Federation notes that as a Party to the Convention the Russian Federation is not included in Annex II to the Convention. 2. The Russian Federation proceeds from the importance of conservation and enhancement of absorbing capacity of forests and other ecosystems, as well as from the necessity of the maximum possible account of this capacity including in the

implementation of the Agreement's mechanisms. 3. The Russian Federation considers unacceptable the use of the Agreement and its mechanisms as tools to create barriers to sustainable social and economic development of the Parties to the Convention."

SOLOMON ISLANDS "... the Government of Solomon Islands declares its understanding that acceptance of the aforesaid Paris Agreement shall in no way constitute a renunciation of any rights under international law concerning State responsibility for the adverse effects of climate change; FURTHER, that the Government of Solomon Islands declares that no provision in this Paris Agreement can be interpreted as derogating from principles of general international law or any claims or rights concerning compensation due to impacts of climate change; AND that the Government of Solomon Islands declares that the low ambition of the Paris Agreement and its adequacy to stabilize global temperature to safe level of below 1.5 degree Celsius, such emissions will have severe impacts and undermining our sustainable development efforts..."

SPAIN

In the case where this Agreement is ratified by the United Kingdom and its application extended to the territory of Gibraltar, Spain wishes to make the following declaration: 1. Gibraltar is a non-autonomous territory whose international relations come under the responsibility of the United Kingdom and which is subject to a decolonisation process in accordance with the relevant decisions and resolutions of the General Assembly of the United Nations. 2. The authorities of Gibraltar have a local character and exercise exclusively internal competences which have their origin and their foundation in the distribution and attribution of competences performed by the United Kingdom in compliance with its internal legislation, in its capacity as sovereign State on which the mentioned nonautonomous territory depends. 3. As a result, the eventual participation of authorities of Gibraltar in the application of this Agreement will be understood as carried out exclusively as part of the internal competences of Gibraltar and cannot be considered to modify in any way what was established in the two previous paragraphs. 4. The application of this Agreement to Gibraltar cannot be interpreted as an recognition of any rights or situations regarding areas not covered by article 10 of the Treaty of Utrecht of 13 July 1713, concluded between the Crowns of Spain and of the United Kingdom.

TUVALU

"The Government of Tuvalu hereby notifies that it will apply the Paris Agreement provisionally as provided for in paragraph 4 of Decision 1/CP.21. [...] The Government of Tuvalu further declares its understanding that acceptance of the aforesaid Paris Agreement and its provisional application shall in no way constitute a renunciation of any rights under international law concerning State responsibility for the adverse effects of climate change and that no provision in the Paris Agreement can be interpreted as derogating from principles of general international law or any claims or rights concerning compensation due to the impacts of climate change. The Government of Tuvalu further declares that, in light of the best available scientific information and assessment on climate change and its impacts, it considers the emissions reduction obligations in the aforesaid Paris Agreement to be inadequate to prevent a global temperature stabilisation level at or above 1.5 degrees Celsius relative to pre-industrial levels and as a consequence, such emissions will have severe implications for our national interests." VANUATU

"WHEREAS the Government of the Republic of Vanuatu declares its understanding that ratification of the Paris Agreement shall in no way constitute a renunciation of any rights under any other laws, including international law, and the communication depositing the Republic's instrument of ratification shall include a declaration to this effect for international record; FURTHERMORE, that the Government of the Republic of Vanuatu declares that, in light of best scientific information and assessment on climate change and its impacts, it considers the emission reduction obligations in Article 3 of the Kyoto Protocol, the Doha Amendment and the aforesaid Paris Agreement to be inadequate to prevent global temperature increase of 1.5 degrees Celsius above pre-Industrial levels and as a consequence, will have severe implications for our national interests..."

Notes: 1 With territorial exclusion in respect of Greenland. See C.N.819.2016.TREATIES-XXVII.7.d of 1 November 2016.

2 For the European Part of the Netherlands.

3 On 13 November 2017, New Zealand notified the Secretary-General of the extention of the application of the Agreement to Tokelau (See CN.705.2017.TREATIESXXVII.7.d of 13 November 2017).

Upon ratification on 4 October 2017, New Zealand notified the Secretary-General of a territorial exclusion in respect of Tokelau (See C.N.723.2016.TREATIES-XXVII.7.d of 4 October 2016.)

4 On 4 November 2019, the Government of the United States of America notified the Secretary-General of its decision to withdraw from the Agreement which shall take effect on 4 November 2020 in accordance with article 28 (1) and (2) of the Agreement. See C.N.575.2019. TREATIES-XXVII.7.d of 4 November 2019.

Appendix B: The Letter to the UN on Withdrawing from the Paris Agreement

UNITED NATIONS NATIONS UNIES

POSTAL ADDRESS—ADRESSE POSTALE UNITED NATIONS, N.Y. 10017
CABLE ADDRESS—ADRESSE TELEGRAPHIQUE UNATIONS NEWYORK

Reference: C.N.464.2017.TREATIES-XXVII.7.d (Depositary Notification)

PARIS AGREEMENT PARIS, 12 DECEMBER 2015

UNITED STATES OF AMERICA: COMMUNICATION

The Secretary-General of the United Nations, acting in his capacity as depositary, communicates the following:

The above action was effected on 4 August 2017.

(Original: English)

"August 4, 2017

The Representative of the United States of America to the United Nations presents her compliments to the Secretary-General of the United Nations.

O. K. Al-Shboul, *The Politics of Climate Change Metaphors in the U.S. Discourse*,
https://doi.org/10.1007/978-3-031-19016-2

This is to inform the Secretary-General, in connection with the Paris Agreement, adopted at Paris on December 12, 2015 ("the Agreement"), that the United States intends to exercise its right to withdraw from the Agreement. Unless the United States identifies suitable terms for reengagement, the United States will submit to the Secretary-General, in accordance with Article 28, paragraph 1 of the Agreement, formal written notification of its withdrawal as soon as it is eligible to do so. Pending the submission of that notification, in the interest of transparency for parties to the Agreement, the United States requests that the Secretary-General inform the parties to the Agreement and the States entitled to become parties to the Agreement of this communication relating to the Agreement.

Attention: Treaty Services of Ministries of Foreign Affairs and of international organizations concerned. Depositary notifications are issued in electronic format only. Depositary notifications are made available to the Permanent Missions to the United Nations in the United Nations Treaty Collection at https://treaties.un.org, under "Depositary Notifications (CNs)". In addition, the Permanent Missions, as well as other interested individuals, can subscribe to receive depositary notifications by e-mail through the Treaty Section's "Automated Subscription Services", which is also available at https://treaties.un.org/Pages/Login.aspx?lang=_en.

- 2- (XXVII.7.d)

The Representative of the United States of America to the United Nations avails herself of the opportunity to renew to the Secretary-General the assurances of her highest consideration.

(signed)
Nikki R. Haley"

***

8 August 2017

Attention: Treaty Services of Ministries of Foreign Affairs and of international organizations concerned. Depositary notifications are issued in electronic format only. Depositary notifications are made available to the

Permanent Missions to the United Nations in the United Nations Treaty Collection at https://treaties.un.org, under "Depositary Notifications (CNs)". In addition, the Permanent Missions, as well as other interested individuals, can subscribe to receive depositary notifications by e-mail through the Treaty Section's "Automated Subscription Services", which is also available at https://treaties.un.org/Pages/Login.aspx?lang=_en.

Appendix C: The Governors' Environmental Activities

The seventeen governors' activities and achievements with regards to protecting the environment are listed in this Appendix (A). This helps explain why these governors in particular stood against the federal government's decision to withdraw the U.S. from the Paris climate accord, with regards to their environmental activities. All of this information below (in Appendix A) was taken from the governors' and states' websites and also from the website of National Governors Association.

1. California

Gov. Brown launched an initiative *America's Pledge* after the president's announcement of his intention to withdraw from the Paris climate deal in 2017. According to him, "America's Pledge brings together private and public sector leaders to ensure the United States remains a world leader in reducing emissions and delivers the country's ambitious climate goals of the Paris Agreement" (https://www.americaspledgeonclimate.com/).

O. K. Al-Shboul, *The Politics of Climate Change Metaphors in the U.S. Discourse*,
https://doi.org/10.1007/978-3-031-19016-2

2. Colorado

Gov. John Hickenlooper worked as a geologist. He announced that he runs for the presidency 2020. He supported many issues that goes against the president's interests, like climate change, gun control, free trader and affordable care act. In 2017 he issued a statement expressing his opposition to the federal government decision to withdraw the U.S. from the Paris agreement.

3. Connecticut

Gov. Malloy signed two climate change bills into law.

"All told, the energy bill I signed today will not only reduce our overall emissions, it will foster growth in the clean energy sector and create good jobs in our green economy". https://www.wshu.org/post/gov-malloy-signs-2-climate-change-bills-law#stream/0.

4. Delaware

Gov. John Carney took the EPA initiative. Tell the EPA Delawareans deserve clean air (https://governor.delaware.gov/clean-air/).

The governor has mad may efforts to protect the environment. Besides to opposing the president's decision, "John was a strong supporter of the Delaware smoking ban and he led the fight for a Cancer Right to Know law, ultimately passed in 2008. John also served as Chairman of the Delaware Healthcare Commission, the Criminal Justice Council, and the Delaware Science & Technology Council" (https://governor.delaware.gov/about-governor-john-carney/).

5. Hawaii

Among Governor David Y. Ige's plans is that he is "moving toward the state's 100% renewable energy goal" (http://governor.hawaii.gov/governors-bio/).

"Hawai`i is a small state with big ambitions and big actions–especially when it comes to addressing climate change. In response to the Paris Agreement, Hawai'i created the Hawai`i Climate Change Mitigation and Adaptation Commission (aka State Climate Commission). The Commission's work focuses on ground transportation emissions reduction and adaptation to sea level rise. As a member of the U.S. Climate Alliance, we believe that smart, coordinated State action can ensure that the U.S. continues to contribute to the world effort to address climate change" http://climate.hawaii.gov/.

6. Maryland

Gov. Larry Hogan "pledged to advance the best ideas, regardless of which side of the political aisle they come from. He is recognized nationally as a strong, independent leader who consistently delivers real results and achieves common sense, bipartisan solutions" https://governor.maryland.gov/governor-larry-hogan/.

"In addition, under the governor's leadership, Maryland continues to be a national and international leader in combating greenhouse gas emissions and has some of the most stringent air quality standards in the nation" https://governor.maryland.gov/governor-larry-hogan/.

7. Massachusetts

"Governor Baker has taken critical and nation-leading steps to diversify the Commonwealth's energy portfolio, safeguard residents, municipalities and businesses from the impacts of climate change, and secure progress toward greenhouse gas reduction targets".

His activities that aims to "lead to half of Massachusetts' electricity being generated by clean resources, while the first ever State Hazard Mitigation and Climate Adaptation Plan stands as a blueprint to prepare for natural hazards and adapt to the impacts of climate change over the next five years" https://www.mass.gov/person/charlie-baker-governor.

8. Minnesota

Gov. Mark Dayton issued statements opposing the federal government's decision (of withdrawing from the Paris climate agreement) in 2017.

9. Montana

"Under the Governor Bullock's leadership, Montana adopted the "blueprint for Montana's Energy Future" in 2016. Written with input from wind energy developers, energy efficiency advocates, small businesses, coal workers, and solar installers, the Governor's Energy Blueprint establishes goals across energy sectors to improve the traditional base of energy generation while sparking a new generation of clean technology business, moving Montana toward more renewable energy, and encouraging innovation, savings, and energy efficiency for homes and for businesses".

Also, he has made important efforts to deal with the impacts of climate change on the state. For example, "Governor Bullock has advanced efforts to better prepare Montanans for climate impacts, ranging from improving long-term drought resilience for Montana's farmer and ranchers through the Governor's Drought & Water Supply Advisory Committee, to advancing forest restoration to reduce wildfire threats through his Forests in Focus 2.0 initiative and work at the Western Governors Association" (http://governor.mt.gov/Issues).

Besides, he encouraged and directed his people to draw on technologies with less gas emission. "His administration has pursued opportunities for new technologies to allow Montana's coal plants to generate energy with less pollution, as well as lowering tax rates for new pollution control equipment, providing tax incentives for EOR, and carbon-capture and sequestration, and supporting legislation that recognized the incidental carbon storage associated with EOR". (http://governor.mt.gov/Issues).

10. New York

"Gov. Cuomo has taken decisive action to lead the fight against climate change and protect our environment for generations to come".

"As the United States now faces an unprecedented reversal of federal climate and energy policy, New York stands ready to work with other regions and states to ensure vital progress on the climate is continued. Governor Cuomo has committed to continue driving New York's environmental leadership and act decisively to oppose any federal action that seeks to roll back progress". https://www.ny.gov/programs/leading-climate-change-protecting-our-environment.

11. Ohio

Gov. John Kasich: According to the state's website, he opposed the President in many issues. He did not support Donald Trumps' running for the president. With regards to climate change, he said

"Now, it does not mean because you pursue a policy of being sensitive to the environment, because we do not know how much humans actually contribute. But it is important we develop renewables. Battery technology can unleash an entirely different world". https://cleanenergy.org/blog/what-presidential-candidates-said-about-climate-change-in-the-florida-debates-this-week/

12. Oregon

Gov. Kate Brown affirmed commitment to fight climate change 1 Jun 2017. In November 12, 2017 Governor Kate Brown Joined by Governors of California and Washington to discuss growth of the clean energy economy.

"Oregon, California, Washington, and British Columbia are members of a regional partnership called the Pacific Coast Collaborative (PCC), which aims to reduce greenhouse gas emissions, decarbonize the economy, and build thriving sustainable communities through the adoption of clean energy policies, sharing of expertise and resources, and collective action". https://www.oregon.gov/newsroom/Pages/NewsDetail.aspx?newsid=2410.

"Governor Brown also signed two executive orders in Portland that move the state toward greater energy efficiency, accelerate the adoption of zero emission vehicles, and lead the state further toward fulfilling goals to reduce greenhouse gas emissions, all while supporting Oregon's economy" https://www.oregon.gov/newsroom/Pages/NewsDetail.aspx?newsid=2402.

13. Pennsylvania

Gov. Tom Wolf. "Tom has worked to protect Pennsylvania's air, water, and land through smarter energy policy and conservation. One of Tom's first actions in office was to reinstate the moratorium on new drilling leases in state forests and parks. He also is working to expand the use of alternative energy like solar and wind". https://www.governor.pa.gov/action-plan-pennsylvania/#ProtecttheEnvironmentandExpandAlternativ eEnergy

"Pennsylvania is the second-largest producer of natural gas in the nation behind Texas," said Governor Wolf. "We are uniquely positioned to be a national leader in addressing climate change while supporting and ensuring responsible energy development, while protecting public health and our environment. https://www.governor.pa.gov/wolf-administration-continues-implementation-methane-reduction-strategy-releasing-new-natural-gas-permits-reduce-air-pollution/.

14. Rhode Island

Gov. Gina Raimondo "The State of Rhode Island is committed to developing practical solutions to climate change through reductions in greenhouse gas emissions, making preparations for the effects of current and future changes in the state's climate, and ensuring RI prospers by being at the leading edge of the transition to a lower carbon future. Below is a list of some of the efforts that the State has engaged in to this end".

Among his contributions to combating climate change, the governor developed a website that aim to "serves as the State of Rhode Island's primary portal for information and resources on climate change and resiliency. It houses resources for a variety of audiences including RI citizens, State agencies, municipalities, non-profit organizations, and the business community. It addresses how Rhode Islanders are working to reduce greenhouse gas emissions as well as how to adapt to the projected effects of climate change such as coastal hazards like sea level rise and storm surge, as well as high heat, drought, and inland flooding" http://climatechange.ri.gov/state-actions/programs-projects.php.

15. Vermont

Gov. Phil Scott signs letter urging U.S. to maintain its commitment to Paris climate agreement. Besides, he issued executive order 12–17, forming the Vermont climate action commission. Among the commission's many responsibilities will be the task of unifying Vermont's ambitious climate and economic goals (https://governor.vermont.gov/press-release/governor-phil-scott-announces-formation-vermont-climate-action-commission).

Change https://governor.vermont.gov/content/preserving-environment

According to the state's website, there are many partnerships & collaborations made by the governor to combat climate change. Among these are:

1. "Joined the U.S. Climate Alliance to demonstrate continued support of the Paris Climate Agreement".
2. "Created, through Executive Order, the Vermont Climate Action Commission to recommend tangible actions to reduce emissions".
3. "Worked with Regional Greenhouse Gas Initiative (RGGI) partner states to agree to lower the CO2 emission cap; Vermont is in the process of adopting the necessary regulations to achieve that result".
4. "Worked with regional partners to update the Zero Emission Vehicle Action Plan to further the electrification of the transportation sector".
5. "Joined the Vermont Climate Pledge Coalition, through which the state is helping track progress toward our emissions goals and encouraging those in Vermont to participate.
6. "Working to Lower Emissions and Increase Energy Efficiency".

7. "Committed to Vermont's goal of achieving 90% renewable energy by 2050".
8. "Committed to investing $2.8 million from the Volkswagen Settlement Fund to the expansion of electric vehicle charging stations in Vermont". (https://governor.vermont.gov/press-release/governor-phil-scott-announces-formation-vermont-climate-action-commission).

16. Virginia

Governor Terry McAuliffe led a number of executive actions:

"On October 31, 2017, Governor Terry McAuliffe …announced the convening of the Commonwealth of Virginia's first Environmental Justice Advisory Council (EJAC). The EJAC, established by Executive Order 73, will provide advice and recommendations to the Executive Branch on ways in which environmental justice should be incorporated in decision-making. Environmental Justice is the principle that no community or individual should bear disproportionate impacts from pollution".

"On June 5, 2017, Governor Terry McAuliffe announced that the Commonwealth of Virginia would join the U.S. Climate Alliance "to move forward on the principles of the Paris Climate Agreement, despite President Donald Trump's decision to withdraw the federal government from the accord".

"Governor Terry McAuliffe signed Executive Directive 11 on May 16, 2017. The directive instructed the Department of Environmental Quality (DEQ) to begin establishing guidelines for regulating CO_2 emissions".

https://zerocarbonvirginia.com/2018/10/07/governor-terry-mcauliffes-actions-on-climate-change/

https://www.governing.com/topics/transportation-infrastructure/tns-virginia-revolving-loan-fund.html

17. Washington

Since he took office in 2013, Gov. Jay Insleehas:

1. "Launched the state's first-ever Clean Energy Fund to support research, and development and deployment in clean energy technologies, smart grid innovation and more. To date, the fund has provided more than $125 million to support transformative projects and create jobs around the state".
2. "Transformed state government operations to reduce greenhouse gas pollution, through to reduce emissions by 14 percent since 2005, thanks to building efficiency upgrades, transitioning the state's fleet to EVs, and transitioning to 100% clean electricity at state government offices".
3. "Used executive authority to limit carbon pollution, by issuing a Clean Air Rule to cap carbon emissions from all the largest sources".
4. "Helped build momentum for state and local climate action through various partnerships such as the Pacific Coast Collaborative, Under2 Coalition, and co-founding the International Ocean Acidification Alliance and the bipartisan U.S. Climate Alliance,—a bipartisan coalition working to uphold U.S. climate goals under the Paris agreement."
5. "Made Washington among the top states for use of renewable energy due to the expansion of solar and wind energy".

https://www.governor.wa.gov/issues/issues/energy-environment
https://www.governor.wa.gov/about/about-inslees/about-jay

Index

O. K. Al-Shboul, *The Politics of Climate Change Metaphors in the U.S. Discourse*,
https://doi.org/10.1007/978-3-031-19016-2

T

GPSR Compliance
The European Union's (EU) General Product Safety Regulation (GPSR) is a set of rules that requires consumer products to be safe and our obligations to ensure this.

If you have any concerns about our products, you can contact us on

ProductSafety@springernature.com

In case Publisher is established outside the EU, the EU authorized representative is:

Springer Nature Customer Service Center GmbH
Europaplatz 3
69115 Heidelberg, Germany

www.ingramcontent.com/pod-product-compliance
Lightning Source LLC
Chambersburg PA
CBHW072053110925
32454CB00003B/10

9783031190155